"[McIntyre] spins the best stories
anyone will ever tell, ever. No o
DOUGLAS W. SMITH, senior wildlif
leader for the Yellowstone Gray W

"*The Rise of Wolf 8* is a saga of
tribal warfare, love and loss, sagacity
insights into the complexity of lup
richly rewarded by reading each de
to a gripping climax to an alto
NORMAN BISHOP, director of the \
coauthor of *Yellowston*

"This book is your invitation an
inside Yellowstone National Park a
more time watching wolves tha
world. As your patient teacher, l
all individuals, and that the
CARL SAFINA, autl
What Animal

"To follow the ever-changir
wolves is to witness a real-l
bravery, tragedy, sacrifice,
monitored the park's
JIM AND JAMIE DUTCHE

Praise for *The Reign of Wolf 21: The Saga of Yellowstone's Legendary Druid Pack,*

winner of the Reading the West Award
for Best Narrative Nonfiction

———

"Wolf lovers, rejoice!"
BOOKLIST

"Like Thomas McNamee, David Mech, Barry Lopez,
and other literary naturalists with an interest in wolf
behavior, McIntyre writes with both elegance and flair,
making complex biology and ethology a pleasure to read.
Fans of wild wolves will eat this one up."
KIRKUS starred review

"Rick's passion for the Yellowstone wolves flows
through this meticulous book about wolf love, play,
life, and death. It's just like being there."
DR. DIANE BOYD, wolf biologist

"Rick McIntyre is a master storyteller and has dedicated his
life to wolves—most particularly Yellowstone wolves. He tells
their stories better than anyone, arguably better than anyone in
history. I too have dedicated my life to wolves, yet reading Rick's
stories, I still learn new things. This book is a treasure."
DOUGLAS W. SMITH, editor of *Yellowstone Wolves:
Science and Discovery in the World's First National Park*

"I'm always eager for the next book by Rick McIntyre.
I learn so much fascinating information about wolves and
their interactions with each other and with their prey."
L. DAVID MECH, author of *The Wolf: The Ecology and
Behavior of an Endangered Species*

"Rick McIntyre has observed wild wolves more than any person ever. It is the way he sees wolves—as fellow social beings with stories to share—that makes his books so powerful. Through that lens, we glimpse our own hopes and dreams."

ED BANGS, former US Fish and Wildlife Service wolf recovery coordinator for the Northern Rockies

"I was skeptical that McIntyre could write a second book as beguiling and insightful as his first about the wolves reintroduced to Yellowstone. Wow, was I wrong. This book is equally captivating as *The Rise of Wolf 8* (which you must read before *21*)."

KAY WOSEWICK, Boswell Book Company (Milwaukee, WI)

"What a tale! It reads like a well-written thriller, except it's real, and you'll never forget Wolf 21 after you've read his story."

LINDA BOND, Auntie's Bookstore (Spokane, WA)

"The continuation of the Yellowstone wolves series did not disappoint. The author did a fantastic job of telling the story of wolf 42 and 21 and the consequences and life of wild animals. I can't wait to see how this series will be concluded."

DAVID OTT, Books-A-Million

Praise for *The Redemption of Wolf 302: From Renegade to Yellowstone Alpha Male*

———

"Rick's books should be mandatory reading for anyone who wants to learn about the fascinating animals with whom we share our magnificent planet and whose lives depend on our good will, decency, respect, along with a deep understanding and appreciation of how vital they are to maintaining a healthy planet in an increasingly human-dominated and troubled world."

MARC BEKOFF, PhD, author of *Rewilding Our Hearts* and *Canine Confidential*

"In this third saga tracing the lives of wolves in the wild, Rick rewards readers with a heartwarming, heartbreaking transition of Wolf 302 from a rambling rake to a responsible adult to a rearguard ready to die to save his pups. Wolf stories don't get better than this."

NORMAN BISHOP, director of the Wolf Recovery Foundation and coauthor of *Yellowstone's Northern Range*

"What thrilling stories about the Yellowstone wolves! The daily observations by Rick McIntyre are unprecedented in their detail and bring the personalities alive."

FRANS DE WAAL, author of *Mama's Last Hug: Animal Emotions and What They Tell Us About Ourselves*

"Patient observer, kind soul, master storyteller, Rick McIntyre is one of a kind. Thousands who have met him in Yellowstone will attest to this. But he has done the most for the wolves; speaking for those who cannot, he has made their stories known. All right here in volume three of an amazing series of books about Yellowstone wolves. There has been nothing else like it."

DOUGLAS W. SMITH, editor of *Yellowstone Wolves: Science and Discovery in the World's First National Park*

"Few people have ever observed wildlife as closely as Rick McIntyre, or written the biographies of individual animals with as much clarity and wisdom."
BEN GOLDFARB, author of *Eager: The Surprising, Secret Life of Beavers and Why They Matter*

"McIntyre's vivid accounts of Yellowstone wolves are reminiscent of Ernest Thompson Seton's animal tales, but these wolves are real and their skillfully told stories offer invaluable insights for researchers and readers alike."
BERND HEINRICH, professor emeritus of biology at the University of Vermont and author of *Mind of the Raven*

"What a gift to humanity is this latest account by Rick McIntyre, the world's most seasoned observer of wild wolves. The book is about birth and death, and living a full life from the perspective of another social species— and it is an unparalleled story of such a life."
ROLF O. PETERSON, author of *The Wolves of Isle Royale*

"Many books have been written 'about wolves.' Rick McIntyre writes about individuals whose lives follow the arc of singular careers. In *The Redemption of Wolf 302* Rick McIntyre continues pushing further into the minds of individual free-living wolves he has known personally for many years. No one else's books have been written with the depth of time, personal insight, and sheer unstoppable devotion that Rick McIntyre brings to the page."
CARL SAFINA, author of *Becoming Wild: How Animal Cultures Raise Families, Create Beauty, and Achieve Peace*

"This engrossing tale of an especially intelligent and charismatic wolf reveals a deep and important lesson: for wolves as for humans there is more than one way to lead a good life."
BARBARA SMUTS, canine researcher and professor emerita at the University of Michigan

Praise for *The Alpha Female Wolf: The Fierce Legacy of Yellowstone's 06*

———

"McIntyre continues his Alpha Wolves of Yellowstone series with this vivid look at the life of the national park's 06 female. McIntyre has a knack for creative comparisons—when 06 kills a grizzly cub to feed her pups, he writes that if she 'were a character in *Game of Thrones*, she would be known as Grizzly Slayer.' Fans of the series will relish this dramatic installment."

PUBLISHERS WEEKLY

"Rick's writing is so vivid, so powerful, that I feel I have been right there with him among the wolves of Yellowstone. And I urge you, the reader, to come with us and discover the magic of wolf society."

DR. JANE GOODALL, DBE, founder of the Jane Goodall Institute and UN Messenger of Peace

"No one has ever had the insight and intuitive understanding of wolves that Rick McIntyre brings to the page, because no one has ever had enough experience with the lifetimes of wild wolves to write their individual biographies. McIntyre is a phenomenon. His series of wolf books is unprecedented in writing about the natural world. And this book's main focus, the career of the wolf called '06,' is astonishing by any measure."

CARL SAFINA, author of *Beyond Words* and *Becoming Wild*

"McIntyre's meticulous observations have gifted legions of observers, and now readers, with new insights into the dynamic lives of Yellowstone wolf packs and with a new empathy for a species that challenges our own worldview, just by being wolves."

ROLF O. PETERSON, author of *The Wolves of Isle Royale*

"The fire and spirit of Yellowstone's matriarch wolves comes alive through Rick's compelling storytelling. The dramatic lives of 06, 42, 926, and other influential females, their loves, losses, hunting prowess, and tender moments of motherhood, unfold in their tenacious fight to survive. These key females leave a legacy that ensures the success of Yellowstone's wolves. Read it!"

DR. DIANE BOYD, wolf biologist

"This tragic story of one of Yellowstone's most beloved wolves is one I know well, thanks to the time I spent in the park with Rick McIntyre and his fellow wolf aficionados. And yet what a pleasure it was to relive all of the joy and excitement she brought wolf watchers during her short time in the spotlight. This book, which places 06 squarely in a lineage of formidable females, will be a treasure for those whose paths crossed with 06's over the years, and a revelation for those coming to her story for the first time. Highly recommended!"

NATE BLAKESLEE, author of the
New York Times bestseller *American Wolf*

"McIntyre's stories let us understand why every hunter/gatherer society in the world admired wolves as 'brothers and sisters in the hunt'—to be understood, respected, and on our most hopeful days, emulated."

ED BANGS, former US Fish and Wildlife Service wolf recovery coordinator for the Northern Rockies

"Rick McIntyre shares gripping and poignant stories about wolves that illustrate their lives in intimate detail. There has never been an adventure story starring wolves comparable to this one."

NORMAN BISHOP, Yellowstone wolf interpreter and adviser to the Rocky Mountain Wolf Project and Living With Wolves

"Rick McIntyre's fourth book in a most remarkable series of books about 'all things wolves' is a gem, as were his first three outstanding books detailing the fascinating and surprising lives of the wolves of Yellowstone."

MARC BEKOFF, University of Colorado, coauthor of *A Dog's World: Imagining the Lives of Dogs in a World Without Humans*

Books in the
Alpha Wolves of Yellowstone Series
by Rick McIntyre

The Rise of Wolf 8:
Witnessing the Triumph of
Yellowstone's Underdog

The Reign of Wolf 21:
The Saga of Yellowstone's
Legendary Druid Pack

The Redemption of Wolf 302:
From Renegade to
Yellowstone Alpha Male

The Alpha Female Wolf:
The Fierce Legacy of
Yellowstone's 06

Thinking Like a Wolf:
Lessons From the
Yellowstone Packs

RICK MCINTYRE

Foreword by **FRANS DE WAAL**

Thinking
LIKE A
Wolf

LESSONS FROM THE
YELLOWSTONE PACKS

GREYSTONE BOOKS
Vancouver/Berkeley/London

Greystone Books Ltd.
greystonebooks.com

Cataloguing data available from Library and Archives Canada
ISBN 978-1-77840-125-1 (cloth)
ISBN 978-1-77840-126-8 (epub)

Editing by Jane Billinghurst
Copyediting by Brian Lynch
Proofreading by Meg Yamamoto
Maps by Kira Cassidy
Jacket and text design by Fiona Siu
Jacket photograph of wolves in the Wapiti Lake pack by John Morrison/
iStock.com. The pack was founded by wolf 755. After he left, his daughter
1091 produced many litters of pups. The pups in this photo could be
grandsons and granddaughters of 755.

Printed and bound in Canada on FSC® certified paper at Friesens. The FSC® label
means that materials used for the product have been responsibly sourced.

Greystone Books thanks the Canada Council for the Arts, the British Columbia
Arts Council, the Province of British Columbia through the Book Publishing Tax
Credit, and the Government of Canada for supporting our publishing activities.

Canada

BRITISH COLUMBIA

BRITISH COLUMBIA ARTS COUNCIL
An agency of the Province of British Columbia

Canada Council for the Arts

Conseil des arts du Canada

MIX
Paper | Supporting responsible forestry
FSC® C016245
FSC
www.fsc.org

Greystone Books gratefully acknowledges the xʷməθkʷəy̓əm (Musqueam),
Sḵwx̱wú7mesh (Squamish), and səlilwətaɬ (Tsleil-Waututh) peoples on
whose land our Vancouver head office is located.

CONTENTS

"A wolf is a living creature, with a perspective, memories of yesterday, an interest in how tomorrow turns out, joys and fears of its own, and a story to be told. These realities create an obligation for us to be concerned for their lives."

JOHN A. VUCETICH,
RESTORING THE BALANCE: WHAT WOLVES TELL US ABOUT OUR RELATIONSHIP WITH NATURE

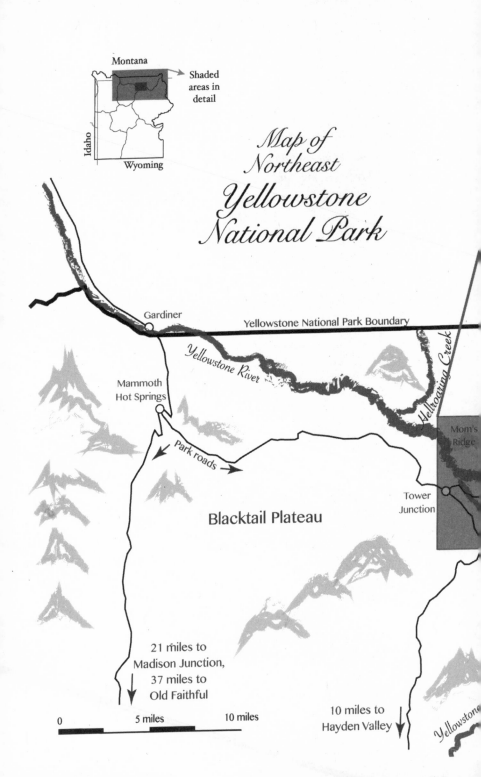

Montana

Shaded areas in detail

Idaho

Wyoming

Map of Northeast Yellowstone National Park

Gardiner

Yellowstone National Park Boundary

Yellowstone River

Hellroaring Creek

Mammoth Hot Springs

Park roads →

Mom's Ridge

Tower Junction

Blacktail Plateau

21 miles to Madison Junction, 37 miles to Old Faithful

10 miles to Hayden Valley ↓

Yellowstone

0   5 miles   10 miles

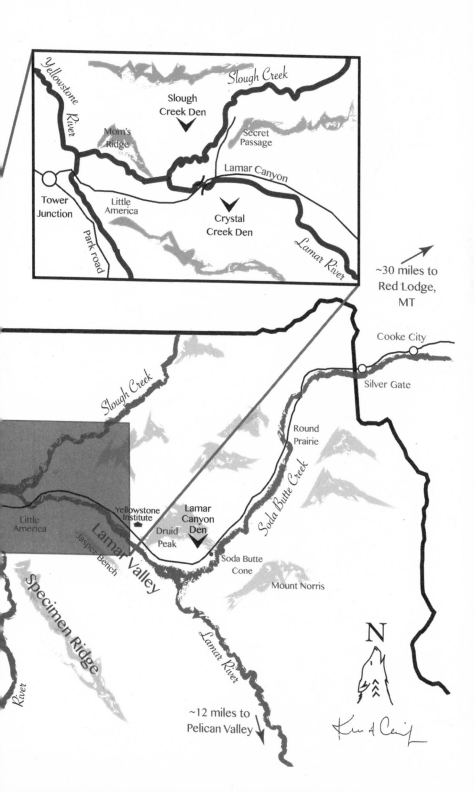

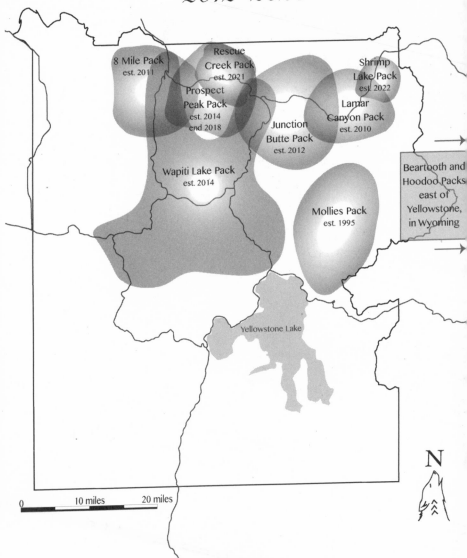

Map of Select
Yellowstone Wolf Pack Territories
2012–2023

8 Mile Pack
est. 2011

Rescue
Creek Pack
est. 2021

Shrimp
Lake Pack
est. 2022

Prospect
Peak Pack
est. 2014
end 2018

Lamar
Canyon Pack
est. 2010

Junction
Butte Pack
est. 2012

Wapiti Lake Pack
est. 2014

Beartooth and
Hoodoo Packs
east of
Yellowstone,
in Wyoming

Mollies Pack
est. 1995

Yellowstone Lake

N

0   10 miles   20 miles

Main Characters 2012–2023

Lamar Canyon Pack

The Lamar Canyon Pack was formed by the 06 Female and brothers 754 and 755. After falling on hard times, it was revived when 926 met 925. After 925's death, 926 recruited four males from the Prospect Peak Pack. Later three males from the Beartooth Pack joined her. Eventually, her daughter Little T took over as alpha female.

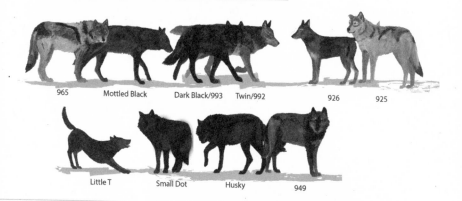

| 965 | Mottled Black | Dark Black/993 | Twin/992 | 926 | 925 |

| Little T | Small Dot | Husky | 949 |

Junction Butte Pack

The Junction Butte Pack was formed by males from the Blacktail Pack and females from Mollies Pack. When all of the original males were gone, males from the 8 Mile and Prospect Peak Packs joined.

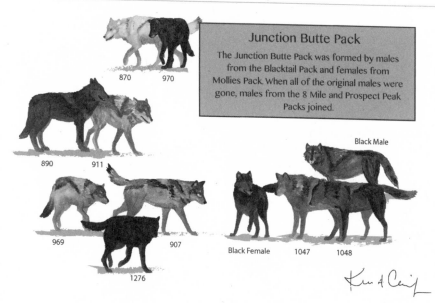

| 870 | 970 |

| 890 | 911 | Black Male |

| 969 | 907 | Black Female | 1047 | 1048 |

| 1276 |

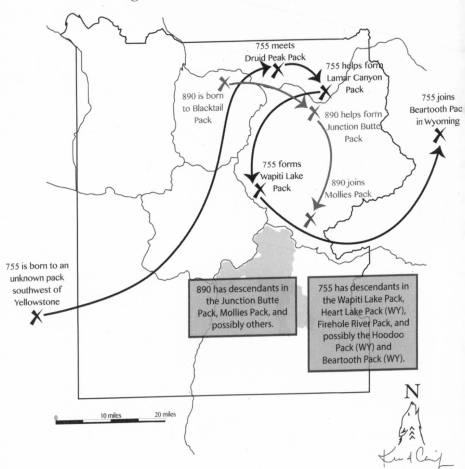

Map of
the Journeys of Dispersers 755 and 890 In and Around
Yellowstone National Park

755 meets
Druid Peak Pack

755 helps form
Lamar Canyon
Pack

890 is born
to Blacktail
Pack

890 helps form
Junction Butte
Pack

755 joins
Beartooth Pac
in Wyoming

755 forms
Wapiti Lake
Pack

890 joins
Mollies Pack

755 is born to an
unknown pack
southwest of
Yellowstone

890 has descendants in
the Junction Butte
Pack, Mollies Pack, and
possibly others.

755 has descendants in
the Wapiti Lake Pack,
Heart Lake Pack (WY),
Firehole River Pack, and
possibly the Hoodoo
Pack (WY) and
Beartooth Pack (WY).

N

0   10 miles   20 miles

PRINCIPAL WOLVES

Dates of birth are included if known.

Lamar Canyon Pack

The Lamar Canyon pack was founded in early 2010 by the 06 Female, who recruited brothers 754 and 755 to be her alpha and beta males.

Alpha females and dates of their reign

06 Female (b. Agate Creek pack 2006)	*2010 to late 2012*
Middle Gray (b. Lamar Canyon pack 2010)	*2012 to 2013*
926 (b. Lamar Canyon pack 2011, daughter of the 06 Female and 755)	*2013 to 2017*
LT (b. Lamar Canyon pack 2014, daughter of 926 and 755)	*2017 to unknown time of her death*

Alpha males and dates of their reign

755 (b. near western border of Yellowstone National Park 2008)	*2010 to early 2013*
925 (b. in unknown pack)	*2013 to 2015*
Twin/992 (b. 8 Mile pack)	*2015 to 2016*
965 (b. Prospect Peak pack)	*2016*
Husky Black (b. Beartooth pack)	*2016 to 2017*
949 (b. Mollies pack, then dispersed to Beartooth pack)	*2017*
SD (b. Beartooth pack)	*2017 to unknown time of his death*

Other Lamar Canyon pack members mentioned

Dark Black/993 (b. 8 Mile pack)

Mottled Black (b. 8 Mile pack)

Junction Butte Pack

The Junction Butte pack was founded in the spring of 2012 by several wolves, including 870 and Puff/911.

Alpha females and dates of their reign(s)

870 (b. Mollies pack)	*2012 to early 2013*
Ragged Tail (b. Mollies pack)	*2013*
870 (second term)	*late 2013 to 2014*
970 (b. Mollies pack)	*2014 to 2016*
907 (b. Junction Butte pack 2013, daughter of Ragged Tail and 911)	*2016 to 2017*
969 (b. Junction Butte pack 2013, daughter of Ragged Tail and 911)	*2017 to 2018*
907 (second term)	*2018 to 2019*
Black Female a.k.a. 1382 (b. Junction Butte pack 2015, daughter of 969 and 911)	*2019 to 2022*
907 (third term)	*2022 to time of writing*

Alpha males and dates of their reign(s)

911 (b. Blacktail pack)	*early 2012 to late 2013*
890 (b. Blacktail pack 2012)	*late 2013 to early 2015*
911 (second term)	*early 2015 to mid-September 2016*
No alpha male from mid-September 2016 through January 2017	
1047 (b. 8 Mile pack 2014)	*February 2017 to 2020*
Black Male (b. Prospect Peak pack)	*2020 to time of writing*

Other Junction Butte pack members mentioned

1048 (b. Prospect Peak pack)

996 (b. Prospect Peak pack 2015)

1109 (b. Junction Butte pack 2015 [?], daughter of 970 and 911)

1276 (b. Junction Butte pack 2018)

Gray Male (b. Junction Butte pack 2018 or 2019)

1229 (b. Junction Butte pack 2019)

Wapiti Lake Pack

The Wapiti Lake pack was founded by 755 and White Female in 2014.

Alpha females and dates of their reign

White Female (b. Canyon pack 2010) *2014 to 2021*

Neck Stripe (b. 2018,
daughter of White Female) *2022 to time of writing*

Alpha males and dates of their reign

755 (b. in Idaho 2008) *2014 to 2016*

Three Mollies males (1014, 1015, and one uncollared
young male) join pack in 2016 and displace 755.

1015 (b. Mollies pack 2014) *2016 to 2018,
then joined 8 Mile pack*

1014 (b. Mollies pack 2013) *2018 to 2021*

1270 (b. 8 Mile pack) *2022 to time of writing*

Other Wapiti Lake pack members mentioned

1091 (b. Wapiti Lake pack 2015,
daughter of 755 and White Female)

1104 (b. Wapiti Lake pack 2017,
daughter of 1091 and granddaughter of 755)

FOREWORD

———

BOTH REVERED AND vilified, the wolf is one of the most misunderstood animals on the planet. Rick McIntyre's eye-opening accounts should be read by anyone interested in the species, whether for positive reasons, which hopefully apply to most readers, or negative ones, such as those who seek to destroy the species. After reading McIntyre's books, a few hunters have even been known to change their minds. Regardless, no reader will ever look at wolves the same way again.

After thousands of days of getting up in the early morning to visit and watch the wolves of Yellowstone National Park, McIntyre knows the toughness of their lives. He deems "resilience" the main trait of these predators as they deal with setbacks from the time they are pups until old age sets in. Their prey resists being brought down and often harms them in the process. They also face competition from grizzlies, who often lay claim to wolf kills. And then there is the continuous rivalry among the wolves themselves, resulting in fierce fights in which they lose territory, their position in the pack, or their lives. About 50 percent of adult wolves in Yellowstone die in fights with other wolves.

On the other hand, there is also great loyalty between pack members, a high level of cooperation when they tackle large prey, and the joy of companionship and playful offspring. Wolves are highly sociable. All of this is described in a straightforward personal style with ample attention to each individual. If there's one thing clear, it's that wolf personalities are as strikingly different as those of humans. No two wolves are alike.

Having extensively documented the power politics of primates, such as in my book *Chimpanzee Politics*, I was most intrigued by McIntyre's descriptions of the wolves' power couples: the alpha male and alpha female. People often forget about the existence of alpha females. Also, in primates (and human society), we tend to focus on male leaders, even though there are always female leaders as well. Chimpanzees are considered male-dominated, but the top female may exert enormous social influence. In wolves, McIntyre believes that the alpha female is the leader of the pack, especially when it comes to decisions that directly impact pack welfare, such as where to den and when and in which direction to travel. The male's duties include keeping the pack safe. And so dominance is contextual: under some circumstances (interpack conflict, the hunt), he acts as the leader, and under other circumstances, she does. People in a long-term relationship won't be surprised by this dynamic.

To start from observations, as McIntyre does, is a technique that is often overlooked in science. Typically, scientists develop hypotheses that they try to support or refute by the data they collect. This approach has its limitations in that hypotheses are limited by the human imagination. They make us ignore unexpected behavior that doesn't fit any existing theory. The more open-ended approach offered here yields a richer picture by just relating what the wolves do daily and how their lives unfold over the years. These accounts will inspire future generations of naturalists by telling them what to look for and what to expect.

There are good reasons to be worried about the wolf's future. This apex predator was virtually exterminated once. Given the fearmongering by its human enemies, this process could be repeated. Providing correct information is critical, such as debunking false stories about how wolves kill more prey than needed, are bad for herbivore populations, or are dangerous to people. Our species must find ways to coexist with wolves since their overall effect on ecosystems is highly favorable. The danger to human life is minimal.

Enjoy the Shakespearean dramas that the Yellowstone wolves engage in as told by one of the greatest experts.

FRANS DE WAAL

INTRODUCTION

———

THIS BOOK IS about what it is like to be a wild wolf and how wolf society works, based on my twenty-nine years of studying wolves in Yellowstone National Park.

Each wolf has its own personality as well as differing levels of physical and social skills, ambition, and willingness to take on risk. Wolves, like humans, are thinking beings that are constantly making choices and decisions. Some decisions are small, such as when to take an afternoon nap or how to hide a piece of meat from your siblings. Others are life-altering.

Yellowstone wolves tend to use four different life strategies during their adult years. Dispersing is the first of those four strategies. For a wolf, dispersing means leaving your family and setting out to find a mate, then forming your own pack.

When I looked up dictionary definitions of *disperse*, the phrase "to spread over a wide area" stood out. In my college biology class, the professor told us that the purpose in life for an individual of any species is to spread its genes as widely as possible. It takes a lot of courage for a wolf to leave its pack and head out into an unknown and dangerous world, because wolves often kill strangers that trespass into their territory.

As in human endeavors, luck can play a big role in how successful a dispersing wolf might be. In the early years of the Yellowstone reintroduction, wolf 21 left his family and wandered into the territory of the Druid wolves a few days after their alpha male had died. 21 was immediately recruited into the pack to replace him. If he had encountered the Druids a few months earlier, he might have been killed by their aggressive alpha male.

Over the years, as I watched dispersing wolves, I tried to figure out what drove them to take on that risk. In the end I concluded that it was the same issue that drives similar human behavior: the ambition to better your situation in life and start a family in a place that offers your sons and daughters the opportunity to thrive. Many of us have ancestors that were successful dispersers. My Scottish ancestors left the Highlands and ended up in Nova Scotia. A later generation moved to Massachusetts. I dispersed from there, worked in several western states and Alaska, and ended up in Yellowstone.

I write about a number of Dispersers in this book, including one I consider the most successful Disperser in Yellowstone history: male wolf 755. I think of those wolves as pioneers and adventurers striking out into unknown lands. By seeking out unrelated wolves to pair off with, they help keep the packs vibrant and strong by avoiding inbreeding.

Dispersers might be called lone wolves, but that term would usually be inaccurate, for dispersing is normally only a temporary stage in a wolf's life. There is always a story when you see a wolf traveling alone, away from its family. If a wolf fails to find a mate and territory, it might return home to its birth pack.

Other wolves choose a less risky strategy. They stay in their pack's territory, accept a subordinate status in life, and hope to eventually replace an alpha when the alpha dies or relinquishes their leadership position because of injury, sickness, or old age. Those wolves could be called Biders, for they bide their time until an alpha position opens up.

Some wolves that remain with their birth packs are more impatient and do not wait for an alpha position to become available. They risk challenging the reigning alpha and aspire to win that title in battle. You could call wolves like them Rebels.

Shakespeare wrote several plays about scheming within royal families to achieve power and status. The series *Game of Thrones* and the many movies about the Mafia deal with the same types of power struggles and betrayals to get to the top. With wolves, you could sum up the strategy of the Rebel with the phrase "To be the alpha you have to beat the alpha." In this book, there are many stories about females in the Junction Butte pack taking over the alpha position by force.

The Disperser, Bider, and Rebel strategies all have the goal of ending up in an alpha position. From then on, the pack will act as a support system for those wolves and their pups.

Then there is the fourth category: Mavericks. These wolves make up their own rules and agendas. Some Mavericks become lone wolves, a category that accounts for only about 2 to 5 percent of Yellowstone's wolf population. Other Mavericks are more social. An example of a Maverick in this book is female 1109, who was a member of the Junction Butte pack but lived life on her own terms, without much interest in climbing the pack's social hierarchy.

As I thought about those issues, I came to realize that in a wolf pack it is good to have members with different dispositions and life strategies so not every wolf is trying to achieve the same things in the same way.

Beyond the life strategy each wolf chooses, there has to be a balance between competition and cooperation within a pack. Although members of a wolf pack compete for status, they must also cooperate when they hunt large prey animals such as elk and bison, when they battle rival packs that have invaded their territory, and when they raise pups. In those endeavors, a wolf pack needs to operate as a well-functioning team.

The continual tension between cooperation and competition is what makes studying wolf social behavior so fascinating, for it so easily relates to human social behavior. All these varied behaviors could be called wolf pack politics, for like human politicians, wolves make alliances and sometimes betray one another.

No matter what life strategies they choose and regardless of what political maneuverings they find themselves swept up in, all the wolves I have studied over the course of my career share one defining characteristic: extraordinary resilience.

The life stories of father and daughter 755 and 926 document the phenomenal resilience of wolves despite terrible losses and setbacks in their lives, while the stories of the Junction Butte pack, especially of father and daughter 911 and 907, illustrate the major life strategies of wolves. Read on for stories of how Yellowstone's wolves deal with traumas, challenges, and tragedies; how they balance cooperation and personal ambition; and how they strategize to achieve what they want out of life. As my Anishinaabe friend John Potter says, "Learn from wolves, not just about them."

1

Dreaming
of Wolves

I WAS GASPING FOR breath after hiking only a few hundred feet and had to stop. My destination was a nearby ridge, a twenty-five-minute hike I had done over a hundred times. The top of that ridge had an expansive view of Blacktail Plateau, home to several wolf packs I had studied over the previous twenty years.

After resting for a few minutes, I continued on, but soon ran out of breath again. That pattern continued until I got to the top. It was July 2015, the middle of our busy wolf denning season, but I realized I had to have my condition checked out. My father died of a heart attack when he was fifty-two, so I was probably genetically predisposed to heart disease.

I went to a cardiologist in Bozeman, Montana. He examined me, then put me on a treadmill for a stress test. The results showed that I had two blocked arteries. The doctor told me I needed to have a bypass operation as soon as possible. I agreed to it right away, for I had a lot to live for, including my studies of wolf behavior and plans to write books about the Yellowstone wolves. I have always been a very optimistic person, so I was confident the operation would go well and bring me back to good health. The operation was scheduled for August 19, thirteen days later, at Billings Clinic, a hospital in the biggest city in Montana. Dr. Scott Milligan, a very experienced surgeon, would be doing the operation.

During those thirteen days, I continued to go out to look for wolves, mainly the Lamar Canyon pack. They were based at a den site in a forest about fifteen miles west of my cabin, which is located just outside the Northeast Entrance to the park in the small town of Silver Gate. I would leave my cabin in early morning, drive into the park, walk off to the south of the road, then watch the den area to the north through my spotting scope. My cardiologist probably would have been horrified if he had seen me hiking like that, but I had to keep monitoring mother wolf 926 and her family.

Most days I got to see 926 and her five new pups, along with other adult wolves in her pack. I had known her since she was a pup and knew her ancestors going all the way back to her great-great-grandmother, wolf 9, who had been brought down from Alberta as part of Yellowstone's 1995 Wolf Reintroduction Project. 926 was descended from a long line of greatly accomplished wolves that included her mother, the o6 Female (so called because she was born in 2006), and her father, wolf 755.

926 was four years old at that time. Yellowstone wolf biologists have found that pups who survive their first winter

have an average life span of only 3.2 years. That means that 926 had already beaten that average.

926 often bedded down at the southern edge of a conifer forest. There was one particular spot near the base of a big Douglas fir tree where she could watch her young pups playing in the meadow below. She could also see a large portion of her family's territory: Lamar Valley and the surrounding mountains. For a mother wolf, those must have been happy times.

As they played, the pups would chase and tackle each other, wrestle, and have tug-of-war contests. Another popular game was hide-and-seek. During one play session, I saw a pup hide behind a tree. It peeked out from behind the trunk for its litter mates, first to the right side of the tree and then to the left, just like a child would do. The pups regularly ran back to their mother and begged for food or play sessions. As I watched the family, I could clearly see that 926 was an attentive, interactive mother.

One day Jim Messina, a Montana man who was deputy chief of staff to then president Barack Obama, joined me to watch 926 and her pups. I filled him in on the successful Yellowstone Wolf Reintroduction Project as well as 926's story. Like everyone who gets to see the park wolves, Jim got very excited. He told me that he would pass on the story of the reintroduction and his sighting of the mother wolf and her pups to the president.

PRIOR TO MY operation, I had gotten up well before dawn every morning for over fifteen years—5,547 days in a row—to watch and study wolves. When the scheduled date for the operation approached, it was hard for me to break that streak, but I had no choice. My friend Laurie Lyman drove me up to Billings on the morning of August 18. The trip to the city took about three and a half hours. I stayed overnight with a friend, then went to the hospital early the next morning.

When I woke up after the operation, I was told that my condition was worse than they had first thought. It turned out that five arteries were blocked. I would have to stay in the hospital for recuperation for at least five days, maybe longer. During those days, I spent a lot of time looking out my window. It gave me a view of the north side of Billings, along with some ridges and cliffs in the distance.

Every time I fell asleep, I had the same dream. I would still be in my hospital room, looking out that window. Everything was the same as in real life, except for one major thing: 926 and the Lamar Canyon wolves were standing out there, north of the city, looking toward me. I understood I was dreaming and knew those wolves were not really there, but I saw them in every dream.

I had already witnessed how 926 had lived a tough life and endured many tragedies. But she never gave up and always moved forward. Seeing her in my dreams motivated me to emulate her determination to advance in life regardless of any setbacks and trauma. She was my role model for dealing with hard times. Those dreams inspired me to recover as quickly as possible so I could get back to the park and see the Lamar wolves in real life. I did everything the medical staff told me to do and kept to a strict regimen of walking up and down the hospital corridors and doing light exercises.

As my release date approached, I was advised to remain in the city for a few more days in case something went wrong with the mended arteries. I stayed at my friend Diann Thompson's house. She was a nurse at the hospital. If I needed to get to the emergency room, it was only a few minutes away. I continued to do all the things I was told. That included going on walks with Diann around her neighborhood. After four days, I went back to the hospital, got checked out by Dr. Milligan, and was cleared to go home.

The rules were that for thirty days I could not drive or carry anything that weighed more than ten pounds. I realized I was going to need a lot of help. A friend in Billings, Ralph Neal, drove me home. Another friend, Kathy Drozda, was vacationing in Silver Gate with her husband, Gregg. Kathy was a nurse and volunteered to watch over me.

Ralph dropped me off at my cabin in midafternoon on August 27. Kathy drove over to check my condition and change my bandages. Later in the day, Darlene Rathmell pulled into my driveway. She and her husband, Ray, had been hosts for many summers at Yellowstone's Pebble Creek Campground. Darlene loaded my spotting scope, my tripod, and a stool into her car, and we drove off into the park to look for wolves. Fifteen miles out, as we approached the area where the Lamar Canyon wolves denned, she pulled over. I got out of the car and she set up my equipment. A few moments later, to the south of the road, I spotted some of the Lamar Canyon wolves, the family that I had seen in all my dreams.

Then I saw the entire pack traveling single file: 926, four adult males, two yearling daughters, and all five of the four-month-old pups. I felt indebted to those wolves. They had come to me in my dreams when I was in the hospital. Now

I was seeing them in real life, as I had hundreds of times before my operation.

FOR THE NEXT ten days, Kathy and Gregg picked me up every morning and drove me out to Lamar Valley. Kathy made me hold a pillow in front of my chest in case we got in an accident. At other times, I joined fellow Wolf Project researcher Lizzie Cato as she checked on wolf packs. Like Kathy, Gregg, and Darlene, Lizzie had to carry my scope and tripod and set them up for me. Jeremy SunderRaj, a wildlife biology student at the University of Montana who I had known since he was eleven years old, would drive down on weekends and take me around.

I was required to go to a rehab center in Livingston, Montana, north of the park, about one hundred miles from my cabin. I had to be taken there many times and Lizzie was my regular driver. At the center, they hooked me up to monitors, then had me work out on various types of exercise equipment. When my readings showed them that I was doing okay, the staff encouraged me to exercise harder. I felt like I was Rocky Balboa training for a title shot.

My doctor instructed me to also exercise at home, so I walked around town and used my NordicTrack when I got back to my cabin. Over the next few weeks, I made a lot of progress toward getting back to normal. I counted off the thirty days and finally completed that required waiting period. Starting on September 19, I could drive myself and carry my equipment when I walked off the road. I felt like a free man again.

By the time a few more weeks went by, I had pretty much gotten back to normal and could do most of the hikes I had previously gone on to better view the wolves. On October 19, exactly two months after my open-heart surgery, I climbed up the ridge that had given me so much trouble before my operation. This time I made it to the top without any difficulty. Then I had to hike up a much higher hill to see wolves and did it easily. I felt better than I had in years.

Thinking back about my operation and recovery period, I was so grateful for all the people who did so much for me: the Bozeman doctor who diagnosed my health problem, my surgeon Dr. Milligan, all the nurses and the rest of the staff at Billings Clinic, especially Diann Thompson. Then there were the people who volunteered to drive me around. I was so lucky to have all those friends.

During my recuperation, I realized that I was in a situation similar to a wolf I had known. Years earlier, when wolves 21 and 42 were the alpha pair of the Druid Peak pack, a yearling male known as wolf 253 was injured while protecting his family from a rival wolf pack.

One of his hind legs was severely bitten and that wound prevented 253 from doing any traveling. He was laid up at Chalcedony Creek in Lamar Valley, a rendezvous site the pack used as a safe place to leave young pups when the adults went hunting. The pups rested and played while the adults were away. Often an older wolf would stay behind to act as babysitter. Chalcedony had a water source and a meadow where the pups could hunt grasshoppers and practice pouncing on voles.

While 253 was recuperating at Chalcedony, other Druid wolves, both pups and adults, regularly visited him. I soon saw that the whole pack was taking care of him. Most days he was left alone when the other wolves went out on hunts. When family members came back from hunts to share food with him, 253 would struggle to stand up so he could greet them. He always seemed calm and fully accepting of his condition.

One day the pack was feeding on an elk carcass west of Chalcedony. A group of Druid wolves left the carcass and headed back to the rendezvous site. The lead wolf was a black pup, and it carried a big elk leg in its mouth. The pup went directly to 253, who was its older brother, wagged its tail, then gave him the elk leg. After that, the pup licked 253 in the face. A few hours later, the alpha pair, wolves 21 and 42, arrived and wolves at the rendezvous site ran to greet them. I saw 253 jump up and run along with them, using only his three good legs.

I kept close track of 253, and on the eighth day after his injury he was doing well enough to accompany his family when they went on a hunt. He gamely kept up with them as he held that wounded leg off the ground. It must have been painful for 253, because he stopped regularly to rest, but he always continued on and caught up with the other wolves. The following month, the pack traveled twenty miles to the west. As I watched the wolves, I saw that 253 was leading at one point. He was still using just three legs, but confidently striding along. 21 passed him but soon 253 limped past his father and took the lead once more.

The way 253's family helped him during his recovery was much like how so many of my friends, neighbors, and coworkers helped me after my operation. All that reinforced a conclusion I had reached many years earlier: the two species on Earth that are most similar in social behavior are wolves and humans.

926 was descended from 21 through her mother, the 06 Female. That meant that 926 was related to 253. I frequently thought of her during my recovery period. The dreams I had of her when I was in the hospital continued to motivate me to have a positive attitude about my rehabilitation, and that enabled me to quickly get back out into the field and resume my wolf research.

One day a thought came to me: Was there something I could do for her? But what could I do for a wild wolf? Years later an incident happened that gave me an opportunity to pay her back.

2

True Grit: 755's Story

WOLF 755'S STORY is intertwined with the story of the legendary 06 Female, which I tell in my book *The Alpha Female Wolf*. The important details to review here are that 06 was born into the Agate Creek pack in 2006. Both her parents (alpha female 472 and alpha male 113) were Dispersers who established a territory at the far western end of Lamar Valley. 06 left her birth family and, in early 2010, formed the Lamar Canyon pack farther east in the valley with young brothers 754 and 755. Even though he was half her age, 755 became her mate and alpha male of the new group.

The brothers were born near the western border of Yellowstone. Like 06 and her parents, the young males had taken the risk of leaving their family and had set out to find mates and their own territory. Their mother and father had both dispersed from their birth packs, so the two young males were carrying on that adventurous tradition.

06 denned at Slough Creek (pronounced *slew*) in the spring of 2010 and had four pups there. The site she chose, known as the Natal Den, had originally been dug out by a coyote family. Female wolves in the Slough Creek pack took over the site in early 2006 and lengthened the tunnel the coyotes had dug. After that denning season was over and the Slough wolves had left the area, a Wolf Project crew measured the den and found that the narrow passageway to the den chamber was fifteen feet long. We figured one fierce female standing just inside the entrance could block an entire enemy pack, for only one wolf at a time could challenge her. By 2010, the Slough Creek pack had ceased to exist and 06, recognizing a good den site when she saw one, moved in.

In 2011, 06 moved her family nine miles farther east in Lamar Valley and had her second and third litters in the same forest where her grandmother, Druid alpha female wolf 40, had given birth to 06's mother eleven years earlier. As told in my previous book *The Alpha Female Wolf*, 755 and his brother 754 heroically protected the den site when a rival pack tried to attack the family. In late 2012, 06 and 754 were tragically shot in a wolf-hunting zone in western Wyoming, just past the eastern border of Yellowstone. Tens of thousands of Yellowstone visitors had seen 06 during her lifetime and they were devastated when they heard of her death.

Father wolf 755 brought the surviving members of the Lamar Canyon pack back to the safety of Lamar Valley. Then, when the February mating season approached, he left to seek out a new mate. It was his second dispersal. He was the father of all the females in the pack, and wolves, like people, have an aversion to pairing off with relatives.

755 took up with a female from the Mollies pack known as wolf 759. Back in the spring of 1996, the ancestors of the Mollies wolves—the Crystal Creek wolves—were denning in Lamar Valley. The Druid wolves attacked their den site and killed the father wolf and all the pups. The two surviving wolves, alpha female 5 and young male 6, were forced to abandon Lamar Valley. They settled in Pelican Valley, about twenty miles to the south. Life in Pelican Valley is hard in winter because deep snow causes elk, normally the main prey of Yellowstone wolves, to migrate to other areas. Bison are much larger than elk and can cope better with heavy snow, so they stay in that valley. The Mollies wolves learned how to successfully hunt bison and survived.

Successive generations of Mollies often made winter trips north to their ancestral home in Lamar Valley. Over the years, they killed a number of wolves that lived in that section of the park, and their descendants regularly fought with descendants of the Druids, including the Lamar Canyon wolves, the pack 755 had just left. The Mollies had a reputation as an aggressive pack, and in his drive to mate with an unrelated female, you might say 755 was taking up with the enemy. But apparently 755 decided finding a mate was more important than carrying on a multigenerational feud.

755 got his new mate pregnant and it looked like she was going to den in Lamar Valley. That turned out to be a mistake, as the Lamar females soon attacked her. Given the packs' history, the Lamar females' aggressive behavior toward a perceived enemy was not surprising. 759 survived the attack and got back with 755, and the two wolves disappeared into a forest on the south side of Lamar Valley. 755 loyally stayed

with the wounded female, but her injuries turned out to be fatal. She was the second mate he had lost in just a few months. Her death meant that the pups within her, pups he had sired, were also dead. 755 was once again without a mate. He lived the rest of the year as a lone wolf, but I had watched wolves long enough to assume that 755's top priority would be to find another female so he could start a new family.

The following mating season, in early 2014, 755 paired off with another dispersing female from the Mollies pack and got her pregnant. When the two wolves briefly left the protection of the park, she was, we think, shot by a hunter. Her wound was serious but not fatal. The two wolves came back to Yellowstone. 755 stayed with her during the two months of her pregnancy and later when she chose a den site west of Lamar Valley. She seemed to be recovering from her wound but soon abandoned her den, a sign that none of her pups had survived. 755 had now lost litters in two consecutive years, a terrible setback, through no fault of his own.

Perhaps because of the stress and trauma they had gone through, the two wolves split up and 755 again became a lone wolf. He lived a solitary life for the next few months. By that time, he was six years old, nearly twice the average life span of a Yellowstone wolf. Then 755 went boldly into what was for him undiscovered country as he sought a new mate.

A FEW MONTHS later, in the summer of 2014, I saw 755 with a young gray female in Hayden Valley, about twenty-five miles south of Lamar. She had been born into the local Canyon pack. Her mother and her grandmother, both famous Yellowstone alpha wolves, had light gray coats that gradually turned white as they aged. The following winter, 755 and his new female mated and the pair became known as the Wapiti Lake pack. *Wapiti* is a Native American word for "elk." 755 and his new mate claimed Hayden Valley as their own, establishing a den site fifteen miles northwest of the Mollies' territory. The female's coat soon started to turn white, making her look like her mother. In our records she became known as White Female.

In the spring of 2015, the Wapiti female gave birth to four pups: two blacks and two grays. It was a happy moment for me when I saw 755 with those pups. I thought of all the tragedy he had endured: the deaths of his brother 754 and his mate 06 in late 2012, the death of his pregnant mate in 2013, and the loss of all his pups in the spring of 2014. His ability to rally and start a new family was, for me, an astonishing example of resilience.

For an old wolf, 755 did well that summer. He killed a bull elk by himself in July. To do that, he had to leap up and grab the bull's throat, then wrestle an animal that was likely six times his weight to the ground. On the way back to feed his pups, 755 was hit by a car. Despite that trauma, he continued to his pups, gave them food from his hunt, rested for a while, then traveled the six miles back to the elk. A grizzly had taken over the carcass, so 755 had to contend with it to feed. Later that month, he killed another elk in the same area. A grizzly tried to steal it but 755 stood his ground and forced the bear to leave. He was fearless when it came to supporting his family.

THE DENNING SEASON of 2015 was very successful for 755 and his mate. All four pups survived through the end of the year. By late February 2016, however, only one pup—a gray female—was with the two adults. Just before that, the family had traveled north from Hayden Valley and wandered across the park border. The loss of the three pups seemed to correspond with that trip and possibly meant they had been shot. That was one more terrible setback for 755. But at least he had that one surviving daughter.

The Wapitis returned to their home base in Hayden Valley, and in the spring of 2016 the mother wolf had four more pups: one black and three grays. With the alpha pair and the 2015 pup, now a yearling, the pack had seven members. Five of them were 755's sons and daughters. Life was back on track for him and times were good for the family.

Everything went well for 755 and his pack through early July. By that time, the mother wolf had moved the three-month-old pups from the den to a nearby rendezvous site. Then a major event radically changed the course of 755's life. On July 7, three big male wolves from the neighboring Mollies pack came into his territory. Three days later, the males chased 755 and his mate, but the Wapiti pair successfully evaded the intruders.

755 had encountered Mollies wolves before. In the spring of 2012, he and his brother 754 had stood up to the pack when sixteen Mollies invaded the 06 Female's den area in Lamar Valley. That day, the two young brothers had saved the family's newborn pups. Now, three years later, 755 was once again having to deal with members of that same aggressive pack.

Things got more complicated on July 11. His mate, White Female, and yearling daughter, 1091, were seen interacting with the intruders when 755 was away on a hunt. The Mollies males were making friendly overtures to the females, and they seemed to be responding in kind. On July 13, people saw 755 back with his four pups and the two adult females at the rendezvous site. The Mollies males were not around and we assumed they had gone back home. We were wrong.

Three days later, the three Mollies males returned: three-year-old 1014, two-year-old 1015, and one young uncollared wolf. 755, at eight years, was an old wolf. He was also significantly smaller than 1014 and 1015. When he was last collared, 755 weighed 88 pounds while 1014 was 128 pounds and 1015 was 123 pounds.

I watched as the three outsiders approached the two Wapiti females. Both females went to the males and flirted with them. All three males did scent marks right in the middle of 755's territory: they appeared to be claiming the area as their own. I spent five hours watching the Wapiti females that day, and at no time did I see 755 or get signals from his radio collar. I did not know if he was intentionally staying away because of the presence of the rival males or just happened to be hunting elsewhere in the territory.

In the following days, we continued to see the three black males with the two Wapiti females and the four pups. It appeared that White Female had accepted the outsiders into her pack. Her yearling daughter, sired by 755, frequently played with the three males. She would be old enough to mate in February and needed an unrelated male to breed with her. The young pups were also getting used to the new wolves as the three males played with them and fed them. This benevolent behavior is typical of male wolves that are new to a pack, and is the opposite of what happens when a male African lion takes over a pride: he usually kills all the cubs born to the previous dominant male. The death of those cubs soon causes the mother lions to come into estrus and the new male mates with them. In contrast, female wolves come into estrus only once a year, in February. By keeping existing pups alive, the new male wolf will also have the potential of breeding with any female pups when they grow up, as they will not be related to him.

755 would visit his pups when the males were gone. One day, he played with the pups for thirty minutes. Another day, he saw a nearby grizzly and monitored it to make sure it stayed away from the pups. That was risky behavior for 755. At the end of July, during one of those visits, the three Mollies males spotted 755 and chased him, with the biggest of the males, 1014, out in front. 755 initially ran from them. When that lead male was about to reach him, 755 turned and charged. The large wolf backed off a bit, then the two males had a standoff. 755 howled defiantly and the rival wolves left him alone. I had known 755 for a long time, so I was happy to see him win that round.

By early August, the Wapiti pups were totally accepting of the three Mollies males. I saw one of the pups go up to wolf 1014 and lick his muzzle, then paw at his face. The big male good-naturedly tolerated the pup's affectionate greeting. Perhaps the pups thought the three males were just additional pack members they had not yet met, rather than unrelated wolves.

I often spotted 755 bedded down where he could watch the family's rendezvous site and his pups from a distance. When the three Mollies wolves left, 755 would trot over to the pups. The big Mollies males gradually grew tolerant of 755, bedding down fifty yards from him without any signs of concern or aggression. I was impressed. 755 had shown his mettle and the Mollies in turn seemed to respect the old wolf.

One morning I saw 755 with his yearling daughter, 1091. They were lying side by side on a low hill, watching the four pups below them. None of the Mollies males was around. 755 and his daughter went down to the pups and lay down among them. The young pups later walked off and the father wolf got up and followed them. When the pups stopped, 755 bedded down once more and continued to watch them. One pup came over and playfully rolled on the ground next to its father. In that moment, 755 seemed totally relaxed and peaceful.

A FEW DAYS after that, I noticed that 755 had a bloody wound on his right hip. I put my spotting scope on him and could see bite marks at the spot. He must have gotten into a fight with one of the males. I had never seen the Mollies males attack 755. In all the incidents I had witnessed, they just chased him off or allowed him to remain nearby. It looked like that tolerance was wearing off. The wounds seemed superficial, but I worried that if 755 continued to stick around he could be injured more seriously.

On August 12, the four pups spotted 755 coming into the rendezvous site and ran over and greeted him affectionately by licking his face. 755 spent an hour with his pups. Then the Wapiti mother wolf and her yearling daughter returned, accompanied by the three Mollies males. The pups left 755 and ran to their mother. Soon after that, I saw 755 leave the meadow. He swam the Yellowstone River and walked off to the west. The time I spent with 755 and his pups that morning was like watching a human father visiting his kids after a divorce, then leaving when his ex-wife and her new husband came home.

That was the last day we saw 755 visiting his pups. It seemed like he had accepted that his ties to White Female and the rest of his family were now over. The final sighting of 755 in Hayden Valley was on August 26.

On September 8, he showed up in Lamar Valley, the place he had once called home. 755 traveled through the south side of the valley. He ended up at Chalcedony Creek, the rendezvous site used in past years by the Druid wolves and more recently by his family, the Lamar Canyon pack. His once-black fur had turned silvery gray, but for an old wolf he looked good.

755 found a bison carcass nearby. I knew that his daughter 926 had recently visited the carcass, and 755 would be sure to detect her scent there. After investigating the area around the carcass, he moved on. We lost him at the east end of Lamar Valley. Soon after that, he disappeared, heading farther east. The Wolf Project's tracking flights in September and October failed to get his signal anywhere in the park. I worried about him. 755 was an old wolf alone in a dangerous world.

WHEN THE THREE Mollies males came into the Wapiti territory and made friendly overtures to White Female as well as to her yearling daughter and four young pups, she had a hard decision to make. Should she side with an old male who might live for only another few years or align with three strong young males? It seemed clear which one of those options would give the mother and her pups a more secure life over the long term. The welfare of her pups was likely her priority. As much as I admired 755, I understood her choice.

After the departure of 755, Mollies wolf 1015 became the next Wapiti alpha male. In winters the Wapitis often come up from their territory in the central section of the park to the northern part of Yellowstone, the area where I study wolves, likely because there are better elk-hunting opportunities there at that time of the year. Two years after the three new males joined the pack, the Wapitis had a skirmish with the 8 Mile pack, who live in the northwestern corner of the park. During the fight, the old 8 Mile alpha male was injured. He left the pack and became a lone wolf.

Soon after that, we saw that the 8 Mile females were traveling with two of the three Mollies wolves that had taken over the Wapiti pack from 755. One of them, 1015, who had been the highest-ranking male in Wapiti, became the 8 Mile pack's alpha male. The third Mollies male, 1014, stayed in the Wapiti pack and became their next alpha male. Given her desertion by two of the three Mollies males, I wondered if the Wapiti mother wolf regretted her decision to choose them over 755.

The last known sighting of 755 in Yellowstone took place on September 21, 2016. I thought that we would never know what happened to him after he disappeared from the park. Then I got a message on November 3. It was from my friend Ron Blanchard, who works for the Wyoming Game and Fish Department in the region east of Yellowstone. He told me he got 755's signal near where his first mate, the 06 Female, had been shot four years earlier.

Ron and I knew that many of 755's daughters were now members of the Hoodoo pack in that area. There was a good chance he would run into some of them or their descendants. They would not be ideal mates for him because of their close family relationship. However, a second wolf pack was based north of the Hoodoos, a group known as the Beartooth wolves. That pack might have a female he could pair off with. We wondered if 755 would pick up their scent. Then Mark Packila, who was doing a wolf tracking flight for the Wyoming Game and Fish Department, reported seeing 755 with two gray wolves in the Beartooth pack's territory. Ron also got 755's signal in that area in early December and later found six sets of wolf tracks where the signals had come from. He felt that 755 was one of those six wolves.

BY MID-DECEMBER WE had a much clearer picture of what was happening with 755. Ron saw him with four black Beartooth wolves. Soon after that, a tracking flight spotted him traveling with ten wolves. The next sighting was even more exciting. Ron saw 755 with Beartooth adults and three of their pups, then watched him play with the wolves in that group. All that seemed to prove he had been accepted into the family.

It must have taken a tremendous amount of courage for 755 to risk approaching that well-established pack. His action reminded me of a phrase I vaguely recalled from my Latin class in high school: *audentes fortuna iuvat* ("fortune favors the brave"). Fortuna was the Roman goddess of luck.

In January 2017, Ron continued to see 755 traveling with the Beartooth pack. An uncollared black was the alpha male. Ron never saw any conflicts between him and 755, which suggested that the other male was totally comfortable with their new member.

I got even bigger news on February 23. It was a text from Ron saying that he had just seen 755 mate with the Beartooth alpha female. What an impressive accomplishment that was for an old male. After that, he and a big male teamed up and killed an elk. That hunt proved that 755 was a valuable asset to the pack.

It soon got harder for Ron to track 755, as the battery in the wolf's radio collar was losing power. A wolf collar battery lasts four to five years, so his collar must have been toward the end of its life span. In May Ron told me that the Beartooth wolves were raising two litters of pups. Since 755 was seen breeding with the alpha female, he was likely the father of some of those pups.

There was one last sighting of 755. In late June, Ron talked to a man who had accidently hiked into the Beartooth pack's rendezvous site. He saw two pups and a collared wolf who was gray but looked like he had originally been black. That was an exact description of 755. At that time, he would have been over nine years old. He had lived nearly three times the average life span of a Yellowstone wolf. That observation of 755 was the perfect conclusion to his life story. He was still doing what he was born to do: sire pups and raise them. After that sighting, he faded away into the wilderness.

THE CHARACTER TRAIT that defined 755 was resilience. Over the nine years of his life, he belonged to four different packs and sired a total of six litters in three different packs. Nothing stopped him as he journeyed through life. There is a saying in the West that is used to describe an especially determined man, one who overcame all obstacles and setbacks in his life and never gave up: *There was no quit in him.* That is how I think of 755.

I got exciting news in the spring of 2017 regarding the Wapiti Lake pack and the legacy of 755. White Female and her two-year-old daughter 1091 both had pups. We later got a count of twelve pups in the two litters. 1091 was sired by 755, so her pups would be his grandchildren. That meant his legacy lived on in the pack. All twelve pups survived to the end of the year.

Wapiti female 1104 was born that year to 1091. She later became a Disperser like her grandfather and traveled far and wide. 1104 went south to Grand Teton National Park, then came up to the northern section of Yellowstone, where I saw her many times. Eventually she and a mate started the Heart Lake pack in the southern section of Yellowstone. 1104 had one litter of pups there in 2020, then moved her family east and claimed a territory outside the park in Wyoming, where she continued to have pups. As of the spring of 2023, she has had eighteen pups. All her pups are descended from 755.

While writing this section, I saw female 1292 from the Beartooth pack in Lamar Valley. 1292 could be descended from 755. 1292 was pregnant when we saw her, so her pups may be additional descendants of 755.

In the early pages of this book, I wrote that the dictionary definition of *disperse* was "to spread over a wide area." That definition serves as a good summary of 755's genetic legacy.

As far as I can tell, he is Yellowstone's most successful Disperser. I do not know of any other wolf that could match his impressive achievement of having pups in so many different packs.

LIKE SOME PEOPLE, wolves such as 755 seem to be especially skilled in knowing how to gain acceptance into a new group. That trait is often called social intelligence. I remember that in my high school years a new guy transferred to my class in the middle of the year. He fit in right away and had plenty of friends from that time to graduation. Social skills seemed to come easily to him. In contrast, when another new boy transferred, he never made friends and eventually switched to another school. 755 would have scored extremely high on a social intelligence evaluation, for his life story showed he knew how to finesse complicated, potentially dangerous encounters with strangers.

Another example of a wolf with exceptional social intelligence is 302, who spent his life in the northern part of the park, from 2000 to 2009. For most of his adult life, he was a Maverick who lived life as he wanted. In his early years, he would come and go from his birth family, the Leopold pack, mate with a number of females in different packs, then return to his parents. The females in his life seemed to forgive him anything, including abandoning them when pregnant. His form of social intelligence enabled him to get away with all sorts of irresponsible behavior for years. Then, when he was an old wolf, he completely changed his behavior. 302 settled down after starting his own pack, the Blacktails, and became a responsible alpha male and father.

A third wolf with superior social skills that gave him an advantage in achieving what he wanted out of life was Druid male 629. In 2007 there was a lot of conflict in Lamar Valley between the Druid wolves and the neighboring Slough Creek pack. Four wolves were killed during fights between the two families. That year, young 629 left home and walked right up to the entire Slough pack by himself.

He carefully avoided their big alpha male, 590, who was the only adult male in the pack, and made friendly contact with the Slough pups and the adult females. Later, when 590 charged at 629, the young Druid ran off—but only far enough to satisfy the other male. After more interactions with other pack members, 629 went up to 590 and licked his face, signifying that he regarded the older male as a higher-ranking wolf. That seemed to satisfy 590, and he left the newcomer alone.

The next day, 629 did the male scent mark known as a raised-leg urination and the Slough alpha female marked over the site. I thought that scent marking by the new arrival would be too much for 590 and that he would attack the stranger and drive him away, but he only briefly pinned 629 and gave him a few minor nips. Soon after that, both males bedded down side by side as though they had been friends for years. It was an impressive accomplishment for the young Druid wolf to become an accepted member of a rival pack, but it looked like he had the right social skills to achieve that result.

In 2022, I cowrote a research paper on scent marking with Dave Mech. I first met Dave, who is considered the world authority on wolves, decades earlier, when I was working in Denali National Park in Alaska. That was when I was just starting to study wild wolves. Dave was very kind to me when I knew nothing about wolves in those early years, and we have been friends and colleagues ever since.

After many decades of research on wolves, Dave still has the excitement level of a young boy when he comes to Yellowstone and watches wolves with us. If a stranger comes along and tells Dave about a wolf sighting, he hangs on their every word. The highest compliment I could pay a wildlife biologist is what I say about Dave: he has the same endless curiosity and quest for understanding nature that Charles Darwin had. He is what all young biologists should aspire to be. Dave contacts me regularly to check if I might have seen some specific wolf behavior he is investigating.

Dave asked if I had observations of young adult wolves in the wild scent marking. In the past, wolf biologists thought only alpha males and alpha females scent marked. I went through my Yellowstone field notes, found many examples of what he was looking for, and sent him those accounts. In some cases, younger adults scent marked sites first marked by the alpha pair. In other situations, alphas marked sites that young wolves had just marked.

I have never seen an alpha attack a lower-ranking wolf who did a scent mark. This had led me to think scent marking is a sign of maturity in young adults, not a challenge to an alpha. That is what seemed to be the case with 590 and 629, even with 629 being a new arrival in the pack.

Both 302 and 629 were successful wolves that used their social intelligence to their advantage, but I would rate 755 as having the highest level of social intelligence of any wolf I have known, based on his success over the years finding female partners, starting packs, and later joining a well-established pack. I have great respect and admiration for him. I would include 755 in the pantheon of Yellowstone's greatest wolves.

3

Living in
the Moment:
926's Story

AFTER 06 AND 754 died in late 2012 and 755 dispersed, most of the remaining Lamar Canyon wolves left Lamar Valley and formed a new group east of the park with some young males from the Hoodoo pack. Since most of the wolves in the pack were Hoodoo wolves, the pack retained that name. That left just two of 755's daughters in the family's territory: 926 and her older sister, Middle Gray.

I remember the time I first saw the pup that grew up to be 926. In the spring of 2011, I spent many hours every morning and evening scanning a forest in the eastern portion of Lamar Valley where the Lamar Canyon wolves were denning. Trees blocked our view of the den, so we had to search open areas for glimpses of the newborn pups.

The previous year, when the 06 Female had denned at Slough Creek, we had a clear view of the den opening. I first saw her pups coming out of the den on May 13, when they were about three weeks old. The mother wanted to get the pups back inside the den, so she tried to push them in with her nose. When that did not work, she picked them up in her mouth, one by one, and carried them back inside.

It took me far longer to see 06's second litter in 2011. On the morning of June 29, I got up at 3:15 a.m. and drove out into the park in the dark. By 5:00 a.m. there was enough light to see the den forest. I parked across the road, walked out to the south, then set up my spotting scope and scanned openings in the trees across the park road to the north.

At 6:06 a.m. I spotted a mob of tiny pups running through a meadow downhill from the den forest. They looked to be about five pounds or so. There were three blacks and two grays. No adults were supervising the pups, who were playing among themselves like kids during recess in a schoolyard. They all ran up on a rock, ran back down, then sped off. As the five pups raced away, they snapped at one another with open jaws. One pup nipped a sibling and that pup nipped back. Soon they ran over the top of a hill and I lost them. It had been just two minutes, but that was the best sighting of the year. I did not know it at the time, but one of the three little blacks was destined to become the wolf we came to know as 926.

A year and a half later, 926's mother and uncle 754 were killed, then her father 755 left to seek out a new mate. I wondered what level of understanding 926 had of those traumatic events. At her young age, how could she comprehend what had happened to her mother or understand why her father had left? All she knew was that the three adult wolves who raised her had disappeared from her life. In the coming years, many more members of her pack would vanish without her knowing why.

THE TERRITORY OF the Lamar Canyon pack was west of the Hoodoo and Beartooth packs, east of the Junction Butte, Prospect Peak, and 8 Mile packs, and north of the Mollies and Wapiti packs. At that time, the Junction pack, the closest neighbors to the two Lamar sisters, had nine members: seven adults and two pups.

To survive as a pack, the two Lamar sisters, 926 and Middle Gray, would have to recruit a dispersing male into their family. That did not take long. A big gray male who came from a pack east of the park showed up in Lamar Valley and joined the small pack. He was later radio-collared and designated wolf 925. That winter he and Middle Gray mated, and in the spring of 2013 she had two pups.

926, now two years old, and 925 worked hard to support the mother wolf and her pups. They would leave the den, go out on hunts, and return with food. Being younger, 926 had been acting subordinate to her sister. But I soon noticed that when male 925 did a raised-leg urination scent mark, 926 would go to the site and mark over it with the female version, known as a flexed-leg urination. I also saw the two young adults playing together and was getting the impression that 925 was more attracted to 926 than to her sister. Perhaps their personalities were just more compatible.

Soon after that, I realized that the female hierarchy had flipped: Middle Gray was now acting subordinate to 926. To be sure this was not a temporary situation while the older sister needed help with her litter, I continued to study the two females. One day the male came in and joined Middle Gray, and the two wolves walked parallel with their shoulders and hips touching. 926 ran over and took control of the situation by squeezing in between the two. Now she was the one touching 925.

The final sign that 926 was the higher-ranking female took place when I saw the older sister go up to her and lick her face submissively. Thinking back over that series of events, I realized that 926 had first formed a strong bond with the male, then dominated her sister when she was still recovering from giving birth. It was a successful example of the Rebel strategy. Later that year, Middle Gray left the pack. She and her one surviving pup were never seen again in the park. We do not know what became of her.

Now there were just two wolves in Lamar Valley: 926 and her mate 925. 926 was the last original Lamar Canyon wolf still in the territory. You could say she inherited the property. If the pack was going to continue to exist, it would be up to her.

926 HAD HER first litter of pups in 2014, at the age of three. She was based at the Druid den forest in Lamar Valley, the site where Druid females had their pups going back to 1997. Her grandmother had been born at the site many years earlier, and more recently the 06 Female had given birth to 926 there. We had a high count of seven pups that spring but soon the number dropped to six. Five of the pups were black and one was gray. Four were males and two females. Those six survived to the end of the year, proving that 926 was a good first-time mother and pack leader.

In March 2015, 926 was in the late stages of her second pregnancy. Undaunted by her condition, she led her pack ten miles west of Lamar Valley, killed an elk by herself in the Junction pack's territory, then brought her family to the site to feed. After finishing off the carcass, the pack began traveling back toward her den to the east.

On the way, the family ran into the Prospect Peak wolves. That pack was much larger and they charged at the Lamars. 926 and the pups immediately turned around and ran off to the west. But the father wolf stood his ground between the Prospects and his fleeing family. The twelve rival wolves reached 925 and attacked him. The battle lasted long enough to enable the mother wolf and pups to get away, but 925 was fatally wounded.

926 was now in a very precarious position. She was a single mother with six existing pups and without a mate to protect and support her. And her due date was rapidly approaching. But she also was a widow who controlled a large territory with a lot of prey animals and a high-quality den site. That meant that suitors would likely come to her. I wondered about the emotional life of a wolf like 926. Specifically, did she mourn the loss of her mate? Based on how dogs seem to be emotionally distressed when close human and canine friends die, I would guess that 926 went through a similar stage of anguish and mourning.

Just a week after Lamar Canyon alpha male 925 died saving his family, mother wolf 926 was faced with another crisis. On March 13, I got 926's signal from the den forest. She was only four weeks away from having her new litter, and her advanced pregnancy would be limiting what she could do physically. Wolf Project biologist Dan Stahler was doing a tracking flight that day. He called down to say twelve Mollies wolves were on top of a high ridge south of the den forest. Dan also saw the six pups from 926's 2014 litter in the den area.

I went back to check on the two packs. People who had been watching the wolves from the roadside told me that 926 had left the den forest and crossed the road to the south. She probably was unaware that the Mollies wolves were up on a ridge a mile or so south of her. Her pups howled from the north and the mother wolf howled back. There were several more bouts of howling back and forth. That alerted the Mollies, longtime enemies of the Lamar wolves, that the smaller pack was nearby. They probably could discern from listening to the howls that there was just one adult.

I spotted the Mollies wolves at the top of a vertical cliff. They were staring toward the den forest. All twelve wolves ran together for a group howl, then looked north again at the den area. For the next four minutes, that pattern continued: a howling bout, looking toward 926's den, then another group howl.

During one of the howling pauses, I heard the six Lamar pups howl. They likely were confused by all the howling to the south, for they knew their mother had gone that way. The Mollies wolves stared toward the sound of the howling pups, then howled again.

I was between the two packs, on the park road, and not too worried. As the Mollies lived in a remote section of the park, they were not comfortable being near a road, especially given the number of cars that had stopped near the den forest. A lot of people were now standing by the roadside to get a view of the wolves. The Mollies were unlikely to try to go through the crowd to get to the pups.

Then the situation changed for the worse. I saw that five of the Lamar pups had crossed the road and were running south toward the sound of the Mollies' howls. They apparently had not yet learned how to tell their own family's howls from the howling of a rival pack.

After wading Soda Butte Creek, the pups stopped and did a loud group howl. When they paused, the pups looked directly toward the cliff where the Mollies were howling back at them. The pups continued running south with raised tails, indicating their excitement at meeting up with what they apparently thought were family members. When the Mollies howled once more, the pups howled in response.

Then I heard howling from a single wolf. It was 926. She was southwest of her pups. The Mollies were southeast of her, still up on the cliff. She switched from a normal howl to a barking call. Wolves rarely bark, but when they do, it warns other pack members of danger. The pups were too young and inexperienced to fully understand her warning. They continued toward the Mollies wolves.

That was the moment the mother wolf mounted a rescue operation, despite the extreme risk involved. I saw her running to where she might be positioned between the pups and the rival pack. Meanwhile, the naive pups were still traveling south, getting closer to the Mollies. They ran into a thick stand of trees and I lost sight of them.

926 now seemed unsure of where her pups were. She stopped and scanned for them, then looked uphill at the Mollies. Likely desperate, she raced on, then stopped and

howled. I heard the pups howl back at their mother. Their calls were coming from a spot between the Mollies and 926. She was now running around with her nose to the ground, frantically trying to pick up the scent trail of her pups.

The pups howled once more and she howled back. I saw that 926 was cocking her head left and right to better determine the exact direction of the howls. She ran off and soon went straight south, nose to the ground, indicating the mother wolf had intercepted the scent trail of the five pups. 926 stopped, looked south, then howled. The Mollies howled back, which must have stressed her out.

I spotted one of her pups in an opening in the trees. A cliff towered fifteen hundred feet above it. The twelve rival wolves were standing at the top. I looked back at the lone pup. Its littermates were in the trees south of it, even closer to the base of the cliff. The pup stared uphill at the wolves there. It looked like it could not figure out how to get up to what it must have thought were family members howling down at it.

I could see that the cliff petered out to the west. If the Mollies noticed that, they could run down and get to the pups, but they did not seem to be aware of that route. The pup could not see that way to the top either, so it stayed where it was.

I spotted the mother wolf north of that lone pup. She stopped and once again tilted her head left and right as she listened to the Mollies' howls and the howls from her pups. I then saw the pup running out of the trees, heading straight toward 926. She spotted it and immediately wagged her tail. Then she ran toward her pup. The other four pups were now in the open and dashing toward their mother, as well. All the tension immediately left me, for now 926 was about to reunite with her litter. The whole family came together and the pups excitedly greeted their mother. She then got them back across the park road and up into the den forest. I looked at the top of the cliff. The Mollies were gone.

The widowed mother wolf, the only adult in her pack, put herself at great risk that day. She had to use her intelligence and experience to figure out what was happening with her pups and a nearby enemy pack. Then 926 ran toward twelve wolves that would have killed her if they had caught her. Ignoring that risk, she successfully gathered up her pups and got them safely back home. It was a heroic effort.

926 and her pups survived the crisis with the Mollies wolves only to face another dire threat. As 926 was getting ready to have her new litter, four of the big Prospect males, wolves who had killed her mate, showed up near her den. I feared they intended to attack 926 along with her nearly twelve-month-old sons and daughters. But then I had a different thought: those four males had no idea of how ferocious 926 could be. Her mother, 06, had dominated all the males in her life, and it turned out her daughter had that same ability.

The next morning, I found 926 with the biggest of those males, a wolf known as Twin, and saw that during the night she had converted him to her side. Then, continuing to take the initiative, she went to the other three males and did the same with them. In one day, she went from being a single mother to having an alpha male and three spare males. Through sheer willpower, she got those four males to serve her. It was a stunning accomplishment, the defining moment of her life. 926 did not let the situation control her; she controlled the situation.

The Prospect males were Twin, 965, Dark Black, and Mottled Black. Twin was later collared and became known as wolf 992. Dark Black was also collared and assigned the number 993. From then on, 926 was the undisputed leader of the pack and those males served her. That included helping her raise her new pups, pups that had been sired by the wolf they had killed.

926 WAS UNDERSIZED for a female wolf but had a big personality. One day I saw the new version of the Lamar pack go to a carcass and noticed that the four males let her feed first. Later the group killed a badger. 926 ate it and did not let the males have a portion until she was full.

A few days after that, 926 and alpha male 992 went to an elk carcass. He stood by as she fed and seemed to be looking around for any threats such as approaching grizzly bears. 926 tore off a large chunk of meat and carried it off toward her den. Rather than feeding, the big male followed her, looking like her personal bodyguard.

On a subsequent day, the five adults were at a big bison carcass but only 926 was feeding. As soon as she finished, the males rushed in and had their turn to eat. At another carcass, only 926 and 992 were present. He must have gotten in her way, for she pinned the huge alpha male and stood over him in the dominant position. 992 weighed 110 pounds when he was collared, while 926 was only 82 pounds.

Then there was the time when I watched the Lamar wolves feeding on an elk carcass and 926 went to one of the big males who was eating a piece of meat and took it from him. He let her steal it. That was more proof 926 was boss of the males. She did not let her size limit her options. Shania Twain's song "That Don't Impress Me Much" summed up 926's attitude toward male wolves who were far bigger than she was.

That spring, 926 turned four years old. She had already proved herself to be a greatly accomplished wolf. Like her mother, 926 was a strong, assertive alpha female who knew what she wanted in life and had the intelligence and willpower to make that a reality. She would need all those assets to get her through many more challenges and setbacks in the coming years.

926'S FOUR SONS soon left the pack, probably because they were uncomfortable with the new adult males. But her two daughters stayed in the family. They were both black and each had a white chest blaze in the shape of a tornado. One of them had a bigger blaze than the other, so we began calling them Big Tornado and Little Tornado. Soon we shortened those names to BT and LT. One of them was destined for great things.

We watched the den forest every morning and evening, hoping to get a glimpse of 926's new litter. I finally saw the pups come out of the den forest in late July. We got a count of five. It was an emotional moment, after all the things 926 had gone through in the past months, to finally see the little wolves running around in a carefree manner.

There was a meadow south of the den forest where the pups liked to play. One morning a big bull bison came into that meadow. He went right up to a gray pup and they stared at each other. The pup seemed confident and unafraid, despite being a few feet from a two-thousand-pound animal. Later that day, two pups played a chasing game and one pup pulled the other one down, making it look like an NFL tackle. Other pups had a tug-of-war over a bone. The bull bison bedded down nearby and two of the pups casually walked by him. They were fearless, just like their mother.

Soon after that, a bison died of natural causes on the east side of the den area. The pack started to feed on it, then a grizzly came on the scene. When it approached some of the pups, one of the female yearlings chased it away. I was impressed by that. It was a classic example of how young adult members of a wolf pack willingly risk their lives to protect the family's pups.

The grizzly soon returned. The adult wolves and pups eventually got used to having a bear feeding just a few feet from them. Later a mother grizzly with two cubs came to the carcass and also ate next to the wolves. 926 seemed skilled at working out understandings with the grizzlies in her area. That was different from the aggressive harassment strategy her mother employed when she felt grizzlies were getting too close to her family.

All five pups seemed healthy. One day I saw them running all out after one of the big Prospect males as they pestered him for a feeding. On another day, male 965 and a pup played together with a stick. A visitor new to Yellowstone watching that interaction would have no way of knowing that none of the four adult males in the pack were related to those pups.

In late August, when I was recovering from my operation in Billings, the Lamar wolves moved their pups across the road to the south. That is where I saw the family, including the five pups, when I got back home. One morning the adults wanted to go farther south but the pups got distracted and began to investigate nearby smells. A yearling grabbed a stick, showed it to the pups, then trotted off. Now obsessed with that stick, all five pups followed. It was a clever way to get the pups to go in the right direction.

That fall mother wolf 926 caught and pinned a stray pup from the neighboring Junction Butte pack. She could easily have killed it. Instead, she nipped the pup a few times, then let it run away. I admired 926 for giving that pup a break. At that time, the Junctions had seven adults and twelve pups, compared with seven adults and five pups in 926's pack. The two adjacent packs mostly kept out of each other's way, and 926 continued what appeared to be a long-standing non-aggression pact between her family and the Junctions.

By that time, 926's pups had developed many skills needed for their survival. If they got separated from the adults, they could find them by following their scent trail. They had also learned how to reunite with the older wolves by going to where they were howling. Despite 926's mothering skills and the help she was getting from the new male recruits, two of the five pups went missing that fall and likely died. The cause of their deaths was unknown. At the end of 2015, the pack numbered seven adults and three pups, an average size for a wolf pack in Yellowstone.

In late January 2016, yearling LT went to Mottled Black and playfully smacked him in the face with a front paw. It looked like she was flirting with him. Soon after that, the Lamar wolves left the park, traveled through Silver Gate, and went farther east, toward the section of Wyoming where 926's mother and uncle had been shot a few years earlier. I always got stressed out when the Lamars visited that area. They soon came back into the park without losing any members to hunters.

One day LT got her face bloody while feeding on a carcass. When walking through a patch of deep powder snow, she lowered her muzzle and plowed through the snow with it. That wiped off most of the blood. After that, seemingly for fun, she dropped down and swam forward through the snow. I had become impressed with LT. She was playful, had a strong personality, and seemed likely to become an alpha female.

By that time, low-ranking male 965 was often away from the pack. When he was absent, that brought the number of adult males down to three. There were also three adult females and three surviving pups. 926 was the undisputed pack leader.

THAT YEAR, MANGE was flaring up in some of the Yellowstone packs, including the Lamars. Mange is caused by mites that live under the skin of canids such as dogs and wolves. Infestations of mites cause irritation and scratching that can lead to secondary bacterial infections. Scratching also results in varying degrees of fur loss. Male 993, the pups, and yearling BT developed bad cases. Their loss of fur made it especially hard for them to survive the frigid winter months. I saw 926 lick a mangy spot on one of her pups and took that to be an attempt to heal the pup. She licked other pups as well. Then I saw alpha male 992 lick mangy areas on 993, his lower-ranking brother. The mange infestation did not affect 992. We never knew why. Perhaps he was born with genetics that gave him better resistance to those parasites.

The patches of bare skin on wolves with mange must have made it painful for them to bed down on snow, so the pack often rested in snow-free spots under big trees. It would be more comfortable to curl up on bare soil. Despite all her skills and abilities, 926 could do nothing to save her pups from mange that winter. All of them died. That must have been a terrible time for her.

It was an awful time for me as well as I watched the suffering of 926's pups. I knew that decades earlier the state of Montana had deliberately spread mange throughout the wolf population by capturing wild wolves, infecting them with mange, then releasing them close to their families so they would infect them. It was a horrible government-sponsored biological warfare program akin to giving Native people smallpox-infected blankets.

In February 926 mated with alpha male 992. The big male seemed devoted to 926 and at times acted like a pup around her. She, however, was not as attached to him as he was to her, for we later saw her mating with lower-ranking Mottled Black. That showed how independent 926 was. She mated with whoever she wanted, regardless of a male's status in the pack. 926 apparently was not attracted to 993, for she rejected his advances. Her daughter LT felt differently. She mated with 993.

There was a disturbing incident that March. While traveling with the pack, 992 collapsed and rolled downhill for no apparent reason. He got up and walked off, but we worried that something was seriously wrong with him. Wolves often get kicked in the head by bison and elk and he may have suffered a concussion or perhaps several of them.

After that, the Lamar wolves traveled west, much farther than ever before, likely because they were not having much luck hunting in Lamar Valley. I drove about thirty miles west and saw them east of the North Entrance of Yellowstone, near the town of Gardiner. To get there, they had to pass through the territories of several other rival packs, including the much larger Junction pack. Although the Lamars and Junctions usually left each other alone, traveling directly through the neighbors' territory was still a risky strategy for the hungry pack. The Lamar wolves survived the trip and all the wolves got safely back home. I attributed that primarily to 926's intelligence and leadership skills. However, I worried that if she continued to trespass into the Junction territory, there could be a fight between the two families.

The Lamars must have been well aware of the risk they were taking, because not long before going on their long trek in search of food, they got involved in a three-way howling contest. 926 and her family were near their den. They looked west intently, like they were hearing other wolves howl, then had a big group howl. We got word that the Junction pack was howling from Slough Creek, a distance of nine miles from the Lamar wolves. I could not hear those howls, but the Lamar wolves apparently could. The Mollies were south of Slough at the time, and they howled from that direction.

BY LATE MARCH 2016, the Lamar Canyon pack was in a much-diminished state. Yearling female BT had left the pack,

as had adult male Mottled Black. That meant the membership was down to just four: 926, her daughter LT, alpha male
992, and beta male 993. But there was reason for optimism.
926 was pregnant once more and would soon have a new litter. Her daughter LT was also getting ready to have her first
litter.

We saw 926's family meeting up with bear 299 at a bull
elk carcass he was controlling. Back then, he was Yellowstone's most famous grizzly and, at twenty-six years old, the
oldest known bear in the park. The wolves tried to dart in and
grab some of the meat. The old grizzly reacted by charging
and swatting a front paw at them. When he got confused
by the number of wolves harassing him, 299 stood up to his
full, intimidating height of nearly eight feet. Apparently, that
did not impress the Lamars. After he dropped down, one of
them bit him on his hind quarters.

The bear had likely dealt with wolves at a carcass many
times and had learned from those encounters. Rather than
chase one wolf off, then another, 299 went to the dead elk
and lay down on top of it. Alpha male 992 went to the far-
thest part of the carcass and fed there, ignoring the big bear
just a few feet away. 926 then did her part: she ran in and bit
the bear on his fat rear end. Soon after that, a compromise
was worked out. The grizzly fed on one end of the carcass
and the wolf pack ate at the opposite end. Yearling LT could
not resist one final act of defiance: she snuck up on the
bear and, imitating her mother, bit him on his undefended
bottom.

That encounter with the grizzly was just a normal part
of life for the Lamar wolves. They had to put up with bears

from spring through late fall, both at sites where animals had died of natural causes and at sites where wolves had done the hard work of making a kill. But in the winter months, when the bears were hibernating, the wolves had the carcasses all to themselves.

IN LATE APRIL, alpha male 922 disappeared, and we never found out what had happened to him. Having seen him collapse for no reason earlier in the year, I suspected a serious head injury. By early May, I could see that both 926 and her daughter LT had signs they were nursing pups. The pack was now down to just three adult members: 926, her daughter LT, and 993, who was the pack's new alpha male. The small pack of three adults had the heavy responsibility of raising two litters of pups that year. Gray male 965 occasionally came back from his travels but did not stay for long.

It was a disastrous spring for 926. At first, she stayed close to her usual den, but we never saw any pups, and after a while she stopped going there. That meant she had lost all her pups two years in a row. We suspected the pups died from distemper, but never knew for sure. 926's once shiny black coat was turning gray, and I wondered if that was partly due to all the stress in her life.

In late June another tragedy hit the pack. Wolf 993 had been alpha male for just a couple of months when his radio collar gave off a mortality signal. My coworker Lizzie Cato and other biologists hiked out to his body. He had a lot of internal hemorrhaging and the Wolf Project crew determined that a moose had trampled him to death. The manner of his death was a prime example of how dangerous it is for wolves, who average about one hundred pounds in their adult years, to challenge elk, who can get up to seven hundred pounds, or moose, who range up to a thousand pounds. About 15 percent of wolf deaths in Yellowstone are caused by prey animals during hunts.

Thinking of all the difficult times 926 had been going through, I was reminded of a saying that goes back to at least 1930, in the darkest days of the Depression: "If it wasn't for bad luck, I wouldn't have any luck at all."

But 926 continued to be defined by her resilience. Soon after the death of 993, I saw 926 playing with her two-year-old daughter LT in what seemed to be a carefree manner. Around that time, I realized that the Lamar wolves were no longer showing signs of the mange infestation from the previous winter. All their coats looked perfect, like they had just been shampooed and blown dry. The small pack had gotten through that terrible time. I believe it was all due to 926's determination to keep moving forward. I remember thinking at the time that character is defined much better by your response to hard times than your behavior in periods of easy living.

926 was down to her last Prospect male: gray male 965, who had recently come back to the pack. She gave him a lot of attention after the loss of 993. They bedded down side by side and did double scent marks as the pack traveled through their territory. But 965 had a pattern of leaving the pack when he felt like it and returning at random times. Technically, he was the alpha male, but I regarded him as unreliable, a Maverick who lived life on his own terms. He reminded me of wolf 302, who also came and went as he pleased. LT was

the third adult in the small pack. I continued to see her and 926 play together frequently.

During one of the mother-daughter play sessions, 926 bumped into 965. That set off a playful chase: he ran off and she pursued him. 926 caught 965 and they took turns jumping up on each other. Her spirits must have been high, for next she ran around in circles, then ran back and forth without anyone chasing her. It was astonishing to see her playing so much and with such abandon, especially after losing all her pups that spring. The pack went back to serious business after the play was over and had a successful elk hunt.

That month I got 926's signal in Silver Gate on several different days. On one of those days, I heard gunshots. It was wolf-hunting season and I worried that one or more pack members had been killed. The following day, the small family was back in Lamar Valley and were all okay.

IN EARLY OCTOBER, four wolves (three blacks and a gray) I did not know chased 926's small pack. Later 965 looked like he had been injured during the interaction. Soon after that, I saw 926 and LT with two new males: a big collared black and an uncollared black. I checked for signals and got wolf 949, a three-year-old male who had been born in the Mollies pack, then dispersed to the Beartooth pack, which was the family 755 had joined late in his life. The other black male had a chest blaze in the shape of a small white dot, so we called him Small Dot or SD. 949 and LT came together and both were wagging their tails. Later LT did a scent mark and SD marked her site. It looked like a relationship was developing between the two young wolves. By that time, 965 had once again disappeared.

A few days after that, another black male joined 926's pack. Since he was friendly with 949 and SD, we figured he was also a Beartooth wolf. He became known as Husky Black. That added up to five wolves in the new version of the Lamar Canyon pack: 926, LT, and the three Beartooth males—949, SD, and Husky Black. For a third time, 926 had recruited dispersing males into her family. First it was 925, then the four Prospect wolves, and now the new group of three Beartooth wolves. That meant there had been eight male adults in 926's life in recent times.

The alpha pair, 926 and 949, did joint scent marks. LT and SD continued to do double scent marks as well. Husky Black also did scent marks with the females. I saw 926 affectionately lick his muzzle. 926 had far more than her fair share of hard times in life, especially regarding losses of male partners. Her luck had now radically changed for the better. With the sudden addition of the three big Beartooth males, both her territory and her family were now far more secure.

926 had a drab-looking coat. If humans had been judging her, she wouldn't have won any trophies at the Westminster Dog Show. But male wolves flocked to her. Was it her feminine allure, I wondered, or the fact that she controlled one of the best wolf territories in the region?

Ron Blanchard, who studies the Beartooth wolves for the Wyoming Game and Fish Department, told me that 949 had been acting as that pack's alpha male since the previous dominant male had been hit by a car in 2015. 949's dispersal from the pack and his obvious interest in the two Lamar females suggested that the females in his pack were all too closely related to him to breed with. If he stayed in the Lamar pack, it would show that finding unrelated females was more important to him than his alpha status in the Beartooth pack. This female issue was also likely the reason the other two males left the pack.

A month after the arrival of those three Beartooth males, we noticed that 949 had developed a bad limp and seemed to be in pain. He lagged well behind the other wolves and often had to stop and lie down. We guessed that he had been kicked in that leg and possibly in his ribs by an elk during a hunt. He moved around better nine days later but by that time he appeared to be subordinate to Husky Black. SD was also deferring to the latest arrival from the Beartooth pack. That sequence put Husky Black in the Bider category, for I never saw him beat up 949. He just seemed to take over the alpha position when the other male's health deteriorated.

Soon after that, a fourth black wolf joined the Lamar pack, a young female, possibly another wolf from the Beartooth pack. She greeted 926 affectionately by pawing at her face. LT was dominant to the new black female. That was the first time I saw an alpha female allow a dispersing female to join her pack. Perhaps 926 let her in because she was strategizing how she was going to compete with her neighbors, the much larger Junction Butte pack.

In late November, the new lineup of six Lamar wolves went out on a hunt and pulled down a deer. They worked well together as a team. We noticed 949 had a fresh wound on his right hip after the hunt. He had probably been struck by one of the deer's hooves.

At times we saw male 965 in the area but he always stayed away from the big new males in the pack. I saw LT go to 965, greet him, then return to her family. 965 eventually gave up any plans he might have had to rejoin the Lamars and was later seen with a group of wolves north of the park.

Husky Black continued to be the highest-ranking male in the pack. SD had become the next highest male in the hierarchy. 949's injuries caused him to drop to the number three position. Despite those wounds and his physical limitations, he was still a well-accepted member of the pack. By late December, he seemed to be feeling better and had resumed doing scent marks. But Husky Black retained the alpha male title.

AT THE BEGINNING of 2017, the Lamar Canyon pack had six black adults: 926, her daughter LT, alpha male Husky Black, 949, SD, and the new female. I was seeing a lot of playing among the pack members, a sign they were all getting along well.

Then something totally unexpected happened. Despite being the top male, Husky Black left the Lamar pack in January, just before the start of the breeding season. That seemed like a strange thing for an alpha male to do, but like people, every wolf is an individual and makes its own choices. A little later, Ron Blanchard told me that he saw Husky Black back with the Beartooth wolves. The recently arrived black female left as well.

Shortly after Husky Black returned to the Beartooth wolves, 755 joined the pack. As I earlier wrote, he was seen mating with the Beartooth alpha female, so some of the 2017 pups probably were his. The prior dispersal of males 949 and SD to the Lamars likely created an opening that may have led to the Beartooths' acceptance of 755.

The departure of Husky Black and the black female from the Lamars meant that the new lineup was 926, 949, SD, and LT. 949 reclaimed the alpha male position. By that time, his leg had healed. I noticed that young male SD played a lot with LT, and I felt his behavior was an indication he was more attracted to her than to 926. Both 949 and SD did scent marks, but 949 was dominant to SD.

IN MID-JANUARY, I found twelve inches of snow on my car and drifts up to twenty-four inches deep in my driveway. I had learned to park my car at the entrance to my driveway at night so there would be less shoveling in the dark predawn hours when I needed to head out to study the wolves. A Park Service snowplow driver cleared the snow from our road early that morning, so I was able to drive out to check on the Lamar and Junction wolves.

926 and 949 mated twice on February 24 and two more times in the next two days. Later LT also mated with 949. We did not see her mate with SD but she easily could have during the night.

I missed seeing a hunt when the alphas took a break from their romantic endeavors, but later I watched video footage of the incident. The wolves approached a bull elk. He seemed lethargic and did not try to run. The video showed 926 leaping up and biting the bull's throat. He lifted her off the ground, but she hung on for some time before she had to let go. The other three wolves attacked the bull's hindquarters. 949 got a good grip on that part of his body. The bull kicked back at the wolf many times. 949 maintained his bite, then let go and tried to attack from a different angle.

The bull frequently lowered his head and tried to gore the wolves with his antlers, but they deftly dodged the potentially fatal thrusts. After a long fight, the four wolves worked together to pull the bull down and then finished him off. It was a classic example of team effort.

That winter I saw 926 do something really impressive. The four Lamar wolves were traveling east near their den site. 926 was in the lead. She suddenly ran forward as though she was chasing something. I looked ahead and saw a mountain lion racing toward a big dead tree. The lion leaped up diagonally and landed on the tree ten feet above the ground. From there she bounded up the trunk and finally stopped about fifty feet from the ground.

926 reached the base of the tree and looked up at the lion. I noticed 926 was alone. Scanning behind her, I saw the three other wolves feeding on a bighorn sheep carcass. The lion must have killed it, but now the wolves had claimed the carcass. 926 turned around and joined the others. When 949 accidentally got too close to 926 as she was feeding, she lunged at him with open jaws. He immediately backed off. The other wolves did that as well and soon 926 had the carcass all to herself.

LT walked over to the tree and bedded down at the base. It looked like she was stationing herself there to block the lion from coming down and returning to her kill. When I checked later, I saw the lion sleeping on a branch.

We eventually determined that the lion had two kittens. That meant a mother wolf had chased a mother lion up the tree. I recalled how 926's great-grandfather, Druid Peak alpha male 21, had a similar aggressive attitude toward mountain lions. 21 was fearless when it came to protecting his family from other predators, and I now saw that his great-granddaughter had that same courage.

IN MARCH THE pregnant 926 attacked a bull elk in a creek and pulled him down. She moved off when cars full of park visitors arrived and parked close to the attack site. We later saw that the bull died of the wounds she inflicted on him. After the people left, the wolves came back and fed at the carcass.

By that time, I could see that LT was also pregnant. Later in the month, when the Lamar wolves did not have any fresh kills, the two hungry females twice went to an old bison carcass where they fed on the tough hide and chewed on bones. After that, even though they were approaching the late stages of their pregnancies, 926 and LT teamed up to chase off a grizzly from a fresher carcass. Later mother and daughter had a play session. Despite their bulging bellies, the two females were very agile as they wrestled and chased each other. In late March, we saw the two males stand back and let the pregnant females feed on a big bison carcass.

Based on when we had seen her mate, we estimated that 926 was going to have her pups on April 29. I did not get her signal that day and took that as an indication she was underground, inside her den.

2017 WAS TO be another tough year for 926. There were no surviving pups that spring, and because they died so young, we assumed once again that they had died of distemper. It was the same story in the Mollies and Junction packs. Because of the twin plagues of mange and distemper, along with unknown causes, 926 had lost all her pups for three years in a row.

Unlike mange, which is caused by mites, distemper is caused by a virus related to the one that causes measles. The distemper virus damages the brain and spinal column, along with respiratory and intestinal functions. It can easily spread from domestic dogs to wild animals, especially wolves and coyotes. Over the years, I had seen many wolf pups die of distemper. Mother wolves who survive distemper can transmit antibodies in their milk to their pups, but that protection ends when their milk dries up. Wolves with black coats seem to have some resistance to distemper that gray wolves lack.

I had never known a wolf who had experienced the loss of so many pups in her life, but 926 did have one adult daughter still in her pack, LT, and it looked like they were closely bonded. That must have helped 926 get though that terrible time. LT had been born into 926's first litter in 2014, so they had been together for over three years. 926 also had adult males 949 and SD at her side.

On August 11, a big bull bison died across the river from 926's den. It was bison mating season and it looked like he had been fatally wounded by another bull. A second bull bison carcass was spotted farther out in the valley. Perhaps that was his opponent. Fights to the death are common among bison bulls.

The following day, I saw alpha male 949 feeding at the bison carcass nearest the den. A huge bison came into the area and headed toward the wolf. 949 stood his ground, even when the two-thousand-pound animal got to within three feet of him. The bison looked at the wolf, then walked off. I noticed that 949 was panting despite moderate temperatures. He also seemed unsteady on his hind legs. After feeding he bedded down by the carcass, in the shade of a willow bush.

The Junction wolves found the second bison carcass farther out in the valley and fed on it. 949 did not move from the carcass he had fed on, and the rival wolves never went near him. Over the next few days, 949 steadily deteriorated. We saw that he licked his fur and paws a lot. People familiar with wildlife diseases told me that his symptoms, including lethargy, indicated he had distemper. 949 had a black coat, so I hoped he would survive.

A coyote came into the carcass site on the fourteenth, saw 949 lying there, and nipped him. The wolf did not seem to have the strength to chase off the coyote. He went further downhill in the following days. When 949 got up and walked a few yards to the nearby Lamar River, he looked drunk. At one point, he stumbled and fell. I could see he was having convulsions in his belly.

By the sixteenth, he had stopped eating and stayed in the shade of the willow bush. When he tried to stand up and walk, he repeatedly fell down. Two days later, he seemed unable to get up from his spot in the shade. A mother grizzly with two cubs fed on the bison carcass but did not bother the incapacitated wolf just a few yards away. Another time I saw five other grizzlies at or near the bison carcass and none of them approached the wolf. Later he tried repeatedly to get a drink from the nearby river but could not stand for more than a moment or two. Whatever strength he had left was rapidly ebbing away.

As I watched 949 during those days, he seemed totally accepting of his lot in life. He stopped trying to get up and just stayed bedded down. His main activity became lifting his head and looking around. A calmness had enveloped him.

One day, six big bison approached where 949 was bedded down under the thick willow. When the lead cow got close, the wolf looked at her and tried to stand up, but failed. I realized that in his helpless state he could be trampled to death if the bison charged. I expected that was about to happen, but the cow stopped, stared at 949 for some time, then turned and walked away. The five other bison followed her lead and left the wolf alone. I have no explanation for why they did not trample the helpless wolf to death.

It ended for 949 on the twenty-fourth. That evening I got a mortality signal from his radio collar, signifying he had not moved in four hours. By that time, I had not seen him eat or drink for nine days. A group of wolf biologists came out from the Wolf Project office the next morning and carried 949 back to the road. Tests later confirmed he died of distemper.

A few days later, I waded across the river to the big willow bush where 949 had bedded down for so many days. The bare soil at the base of the bush was wet, indicating there recently had been standing water there. The wolf might have been able to lap up some of the moisture. It was shady and cool under the willow's dense branches, which probably provided 949 with some relief from the hot August temperatures in the valley. From his spot under the willow, 949 could have lifted his head and seen much of his territory. Other Lamar wolves had been spotted close to his location, so 949 could have heard them howling. But I never heard him howl during the nine days he was rapidly declining. I wondered why that was.

Distemper is highly contagious and can easily spread when wolves come into direct contact, which frequently happens when packmates greet each other. Did the other Lamar wolves have some understanding of the danger of getting close to a sick wolf? If so, that might explain why none of them visited 949 during his last days. Then there was a more profound question: Did 949 sense that what was making him sick could affect his packmates, and was that why he remained silent as his health declined?

The death of 949 was one more terrible upheaval, one more loss in 926's life. But there was a replacement male in the pack. Young SD ascended to the Lamar Canyon alpha male position. He was the smallest of the males who had recently joined the pack and now was the last male standing in the family after the deaths of 925, 949, and 993 and the dispersal or disappearance of 965, 992, Mottled Black, and Husky Black. SD was destined to stick with the Lamar wolves through thick and thin for the rest of 926's life. Her relationship with him was to be the longest of her life.

THAT FALL, I frequently got 926's signal outside Yellowstone. One evening in September, I saw her in our small town of Silver Gate. The wolf-hunting season was about to start and I was worried about her safety. By late September, she was back in Lamar Valley with her daughter and SD. Things seemed like they had returned to normal in the pack. I saw 926 and SD playing together. They looked like exuberant young pups who had no worries in life. All three adults scent marked the same spot. They hunted together and fed on old carcasses when nothing else was available.

Seeing 926 play with the much younger male wolf in such a carefree manner brought to mind "Invictus," the famous 1875 poem by Englishman William Ernest Henley. In his early adult years, Henley suffered from many medical issues and had a leg amputated. Doctors just barely saved his other leg. Henley wrote the poem during his recovery.

Invictus is Latin for "unconquered." The poem praises resilience in the face of adversity. It stresses an individual's power to control responses to hard times. Winston Churchill frequently quoted the poem in the darkest days of World War II. Nelson Mandela often read "Invictus" to fellow inmates in the Robben Island prison.

> Out of the night that covers me,
> Black as the Pit from pole to pole,
> I thank whatever gods may be
> For my unconquerable soul.
>
> In the fell clutch of circumstance
> I have not winced nor cried aloud.
> Under the bludgeonings of chance
> My head is bloody, but unbowed.
>
> Beyond this place of wrath and tears
> Looms but the Horror of the shade,
> And yet the menace of the years
> Finds, and shall find, me unafraid.

It matters not how strait the gate,
How charged with punishments the scroll,
I am the master of my fate:
I am the captain of my soul.

The poem effectively sums up what I imagine was 926's attitude to all the tragedies in her life. Her head was unbowed and she was the master of her fate. If any wolf had an unconquerable soul, it was her.

THAT FALL I confirmed something I had suspected. LT was the alpha female and 926 was not disputing the issue. It looked like 926, who was now an old wolf, accepted that it was time for her daughter to take over. A peaceful transfer of power to the next generation had taken place. 926 was a Rebel, for she had taken the alpha female position away from her older sister. LT, in contrast, was a Bider. Her life had been devoted to helping her mother. From now on, 926's job was to support her daughter.

On December 1, 2017, the pack traveled to where a bull bison had died eight months earlier. There was not much left on the carcass. At one point, both 926 and SD were trying to get tidbits from opposite sides of the big skull. SD later left and LT took his spot. She and her mother continued to work the skull for tiny scraps. Later that month, the three Lamar wolves continued to travel out of the park to Silver Gate but always came back to the safety of the park.

On the last day of the year, I saw 926 flirting and playing with SD, who was likely less than half her age. He scent marked a site and she marked over it. The pack had a new elk kill that day. I saw the male carry off a piece of meat and give it to 926. Seeing him do that for the old wolf was an emotional moment for me. That play session and the gifting of the meat became one of my favorite memories of 926.

The Yellowstone Wolf Project's lead biologist, Doug Smith, did a wolf tracking flight a few days after I had watched the Lamar wolves feed on that elk carcass. That morning I had gotten signals from 926's family but could not spot them. I watched as the plane circled a high ridge. The wolves had to be there but I still could not find them. I texted Doug and asked where they were. A minute or so later, I got his detailed reply and I quickly located them sleeping on a high ridge. I would not have seen them without his help. Doug is modest about his wolf spotting abilities. Whenever I compliment him on that, he tends to say, "Even a blind pig sometimes finds an acorn."

In late 2017, the Lamar Canyon pack was once again at a low ebb. There were only three members: 926, her daughter LT, and alpha male SD. They were still scavenging on the eight-month-old bison skull. Times continued to be hard for 926 and her family.

AT THE START of 2018, the Lamar Canyon pack was the smallest of the eleven wolf packs in Yellowstone. That contrasted with two neighboring packs: the Junctions to the east had eight members and the Mollies to the south had fourteen. The Wapitis in Hayden Valley were now the largest pack in the park. They numbered twenty-one. The Junctions and Mollies wolves were not making significant incursions into Lamar Valley, so 926's family was holding their territory for now. But that could change at any time.

At their peak, the Druids had been the most dominant pack in Yellowstone. They numbered as many as thirty-eight wolves and controlled a huge section of the northern part of the park, from the eastern border through Lamar Valley, and as far west as Hellroaring Creek, a span of forty miles. The three Lamars had a much smaller territory: the highly valued Lamar Valley and less productive forested areas east of there.

On the second day of January, I spotted Lamar wolves SD and LT playfully sliding down a snowy slope. After that, 926 affectionately licked the face of the male. In February alpha pair LT and SD did a double scent mark, another sign that they were the alpha pair. 926 did not mark the site.

During the February mating season, I saw SD mate with LT. Later both females were pregnant, so he must have also bred with 926. Mange and distemper had subsided in the park, so it looked like this was going to be a good year for the family.

SD still seemed to be more emotionally attached to young LT than to 926, who was much older than him. However, one day that winter I saw 926 and SD playing together in a happy-go-lucky manner. 926 would soon be seven years old and her once-black coat was now turning silver.

Around that time, a Wolf Project tracking flight located the Lamar wolves in their den forest. They were digging out an old bison carcass that had been covered with snow, presumably because there was nothing better for them to feed on.

THAT MONTH, THE Wolf Project planned to use a helicopter crew to capture and radio-collar LT and SD, the Lamar alpha pair. We found the pack a few miles east of their den. They had just traveled into a dense forest. I radioed the crew and directed them to come to our location.

When adult Yellowstone wolves hear a helicopter heading toward them, they run away. They have learned that sound means humans are trying to capture them. The younger wolves follow their example and flee with them. The helicopter flies alongside the running wolves as a park biologist shoots a tranquilizer dart into each of the targeted wolves. Then the helicopter lands so the crew can radio-collar the wolves. At least this is how it is supposed to work. But that day, 926 stayed in the trees when she heard the helicopter approach. It was the ideal strategy to foil a capture operation.

The helicopter flew back and forth but 926 did not budge. I felt that showed her intelligence, for she must have figured out that response based on past collaring attempts. The other two wolves apparently trusted her judgment, for they stayed put as well, even though their most basic instinct would have been to flee.

NO MATTER HOW large or small they are, wolf packs benefit from having older members who can show younger, inexperienced wolves what to do in new situations and dangerous encounters with rival packs and prey animals. Because of the openness of the Yellowstone landscape, and decades of observations by biologists and the cadre of local people and repeat visitors known as wolf watchers, we have extensive case histories of fights between rival wolf packs. My Wolf Project colleague Kira Cassidy and her coauthors Dan MacNulty,

Dan Stahler, Doug Smith, and Dave Mech published a research paper that analyzed the factors that tend to give one pack a victory over another. The title of their 2015 paper was "Group composition effects on aggressive interpack interactions of gray wolves in Yellowstone National Park."

The paper first reviewed our records and stated the expected fact that larger packs tend to win battles with smaller packs. For example, the odds of a pack defeating a rival group when both have an equal number of members is even (1 to 1). The authors found that having just one more member increased a pack's chances to 2.4 to 1. If a pack is 20 percent larger than the opposing pack, the probability of winning jumps up to 80 percent, compared with 50 percent if the sides are even.

For me, the most intriguing finding of their research is the value of having old wolves in a pack. Kira went through Wolf Project records and found that when packs have an equal number of members, having a senior wolf in the family, six years or older, greatly enhances the chances of victory for that pack. As the authors stated, "This old wolf effect is probably due to the elders' years of fighting experience. They have encountered many opponents, have killed rivals, and have seen pack members killed. They have a steadying influence on their packmates during the early, chaotic moments of the fight as they divide the opponents, preventing them from regrouping, and eventually driving them away." Seven-year-old 926 fit that profile.

I would compare that finding to what happens in human warfare when an experienced sergeant takes a platoon of new recruits into battle. The young soldiers will look to their

sergeant when they come under fire and likely will do whatever he or she role-models.

After reading Kira's paper, I recalled an incident in the
early years of the Wolf Reintroduction Project when wolf 8
was the alpha male of the Rose Creek pack. He was out hunting with several inexperienced yearlings when the Druid
pack invaded their territory and charged at them. The Druid
alpha male, wolf 38, was much bigger than 8. But without
hesitation 8 ran toward the opposing alpha. Every yearling in
his group immediately followed his example. 8 reached 38,
attacked him, and pulled him down. The younger wolves ran
in and joined 8 in biting the big male. After beating 38 up, 8
stepped back and let his opponent go. The other Druids saw
that their alpha male had been defeated and ran off.

Another example of the value of older wolves took place
when ten Mollies wolves came into the territory of the
Druid Peak pack in 2007. Druid alpha male 480 immediately
marched toward the rival pack. He had only five pups with
him, but they bravely followed 480 and helped him attack
the larger force. Because of 480's brilliant leadership and the
support of those pups, their group of six easily defeated ten
opponents.

A third case happened a few years later. It involved Druid
adult female 571, who had been one of the pups that followed her father 480 into battle. When the other adults in
her family left on a hunt, she stayed behind to watch over the
family's pups. A yearling male known as Triangle was helping
her. Three big males from a distant pack suddenly showed up
in Lamar Valley. I saw that they were heading directly to the
pups' location.

571 risked her life by running toward the Druid pups and the invading wolves. Triangle followed his older sister into the dangerous situation. The three males saw them coming and charged at the two Druids. That drew the intruding wolves away from the vulnerable pups. One of the invaders caught 571 and attacked her. At great risk to his own life, young Triangle ran in and bit him. That enabled 571 to escape. All the Druids survived thanks to 571's courageous leadership and the support her little brother gave her.

I think what happens in these cases is that when younger pack members see older adults take decisive action in a crisis involving a rival pack, it overcomes their natural feelings of panic and fear, giving them confidence to charge forward with those adults.

The same thing likely happens when pups start to accompany adult pack members on hunts. They have to override their fear of attacking prey animals much bigger than they are. That naturally happens when they see older wolves biting at a targeted elk or bison. They join in, probably without stopping to think about the danger of being hurt or killed.

The conclusions in the Wolf Project paper also related to an incident I witnessed back in November 2015, when sixteen wolves from the Mollies pack traveled to Lamar Valley from their territory to the south. We spotted the Mollies chasing Lamar male Mottled Black. They caught and attacked him, but he managed to escape. I could see that he was wounded, but not fatally. Two days later, 926 led her pack to a place where the Mollies had been. She did a scent mark there, which I took to be her way of showing defiance to the rival pack. Then two of the big males in the family and a yearling daughter followed her example and also did scent marks there.

I concluded that having more than just the two alpha wolves scent mark could work as a deterrent to other wolf packs that might otherwise invade the territory. If the Mollies wolves came back to where the four Lamar wolves had scent marked the territorial boundary, they would quicky figure out that there were multiple mature adults in the Lamar pack, which could cause them to leave the territory without having a confrontation.

Multiple scent marking by wolves is essentially an advertisement of the pack's strength. The behavior warns potential invading packs that the territory is guarded by numerous older adults. It does not matter if we are thinking about wolf packs or human armies: the best victories are the ones you don't have to fight.

The value of having multiple older wolves in a pack might also work to moderate the aggression of alphas. If an alpha wolf drove off middle-aged or older wolves, its pack might lose a battle with a rival group. It would be in the interest of alpha wolves to keep an adequate number of adults in the group so that their side could prevail.

That thought took me back to the story of Druid male 629, who was unexpectedly accepted into the rival Slough Creek pack. Slough alpha male 590 was the only adult male in the pack at the time. By allowing the younger male to join his pack, 590 was increasing the chances that the Slough wolves could win fights against rival packs. In addition, now two adult males in the pack were scent marking, which might make other packs more likely to avoid a confrontation with the Slough wolves.

THERE ARE DOWNSIDES to getting old when you are a wolf. By the end of January 2018, I noticed that the two younger Lamar wolves would often be well ahead of 926 as the pack traveled. That worried me, for it indicated 926 was having trouble keeping up with them. As she always caught up eventually, I wondered if the other two wolves slowed down a bit in deference to her experience in hunting and other endeavors. As I watched her during those times, I was reminded of the phrase I had applied to her father, wolf 755. Like him, there was no quit in 926.

I wanted 926 to live many more years but the truth was that she was likely nearing the end of her life. My hope was that 926 would have one last litter that spring and successfully raise a batch of healthy pups, both her own and her grandpups, to the end of the year. After all the hard times she had lived through, especially the loss of all her pups in the previous three years, 926 deserved a good year.

I finished up my Park Service job with the Wolf Project at the end of February 2018. I continued to go out early every morning to study the wolves, then worked in the afternoons on writing books about my experiences with wolves in Yellowstone.

In early spring, we saw that the two Lamar females looked pregnant. By April, they were both staying close to the Druid den forest. We later noticed that 926 was nursing, proof she had pups. We felt LT was also a mother. By that time, 926's radio collar had fallen off. As the attempt to collar the alpha male had been unsuccessful—thanks to 926's tactic of staying well hidden in a stand of trees instead of making a run for it— that meant that no wolves in the pack were collared, which made it harder for us to find them and follow their movements.

I got a good look at 926 in early June. She had a full belly and her coat looked perfect. She carried a piece of meat into the den forest, likely to give to the pups. On a later day, 926 brought part of a deer carcass up to the den area. Laurie Lyman saw 926 get a pronghorn fawn in mid-June. She was no longer the leader of the pack, but the old wolf was working hard to support what we assumed was two litters of pups.

We never did see the pups while they were in the den forest. It was not until October that year that I finally spotted them. There were four and all were black. They gave 926 a lot of attention. Some of the pups were likely hers, while others would be her grandsons and granddaughters. I imagined that these were very happy times for 926. She was doing what she was born to do: raising a new batch of pups. The defining characteristic of 926 was her resilience. To steal a term from the *Star Wars* movies: the force was strong with her. She seemed determined to move forward regardless of how many losses she encountered. Nothing could stop her. 926 always chose to do something about whatever was holding her back.

926's radio collar was later found in her den after the family left the site. It looked like the leather strap had been chewed through. 926 could not have done that herself. Like dog puppies, wolf pups like to chew on things. I guessed 926's pups chewed through her collar when it was around her neck. It was endearing to picture her good-naturedly tolerating the pups doing that.

Yellowstone's biologists normally radio-collar only about 25 percent of the park wolves. They try to have at least two collared wolves in each pack for research purposes, and we usually have about ten packs in the park. Some collars are equipped with GPS capability so that pack movements can be tracked by satellites.

In November 2018, the Lamar Canyon family often traveled to Silver Gate. Six of them were seen about a mile south of my cabin, on a back road called Bannock Trail. By that time, one of the four pups was missing. The location of the pack was a problem, for the Montana wolf-hunting season had opened and Bannock Trail was in Montana.

On the twenty-second, I heard that 926 had been spotted just off the road between the towns of Silver Gate and Cooke City. Two other black wolves were nearby. Lots of people had gathered and one man was talking about shooting wolves. That guy drove back to Cooke. Soon two vehicles arrived from there and the people in them made anti-wolf comments. Our area was a dangerous place for 926 and her family.

It was not unusual for me to hear gunshots when I was in my cabin, but I never knew if someone was shooting at an animal or doing target practice. I do not remember if I heard shots on the afternoon of November 24.

We soon found out that a wolf had been shot and killed on Bannock Trail that day. It turned out to be 926. Word immediately spread on social media and throughout the world. People who had known 926 or heard about her were devastated. Our only solace was that her adult daughter, the family's three pups, and the father wolf had all survived.

As when 926's mother, the 06 Female, was killed in late 2012, I tried to comfort people who were grief-stricken by the loss of a wolf they greatly admired. Concentrating on that helped me cope with the tragedy of her death.

At the time, there was a big controversy about whether 926 had been shot legally. A game warden from the Montana Department of Fish, Wildlife and Parks came to town to investigate. He interviewed the shooter, who was not a Silver Gate resident, and witnesses. A neighbor called me to talk about the shooting and investigation. We decided to go to the section of Bannock Trail where the incident took place. When we arrived, I saw a large patch of blood-soaked snow on the road. We had heard that 926 had been shot off the road, dragged to this spot, loaded onto an ATV, and taken away.

I stared at the red snow for a long time. Given the copious amount of blood, it looked like 926 was alive when she was brought here, then bled out at the site. She had been allotted a life span of seven years and seven months. That is more than twice the average life span of a wolf in Yellowstone National Park. As we stood there, I knew the investigation was still ongoing. It occurred to me that it would be a good idea to take a sample of the blood in case it was needed. I scooped up a good quantity of the red snow, put it in a sturdy plastic bag, and left. When I got back to my cabin, I had to decide what to do with the bag. I did not want to leave it unattended outside, so I put it in my freezer, intending to keep it there until I knew if it was needed for evidence.

My last sighting of 926 had been thirteen days earlier, in a meadow about ten miles west of my cabin. I was in a small parking lot, watching her from my car. She came to the road and walked past my vehicle. I waited until she veered off to the north, then drove to another location. From there I could see 926's adult daughter LT and the family's three pups. They had a greeting ceremony. 926 joined the group and the family moved off to the north. I lost them in trees but soon saw 926 farther to the east. A few minutes later, the others followed her route. Then they all went into a forest and I lost sight of them.

There is one more 926 story to tell that has a more positive and respectful ending than what happened to her on that back road in Silver Gate. I will tell that story at the end of the book.

4

Rivals:
970's Story

IN THE SPRING of 2012, we discovered that a new group of wolves was denning in the Antelope Creek area, southwest of Lamar Valley. It turned out that several young sisters from the Mollies pack had dispersed as a group and met up with some dispersing males from the Blacktail pack. The ancestors of those Mollies and Blacktail wolves had feuded with one another for generations, but this new group of young wolves put all that bad blood aside. We called them the Junction Butte pack.

By early 2013, the pack had seven adults and two surviving pups. The gray alpha female was a former Mollies known as wolf 870. The gray alpha male, from the Blacktail pack, was originally known as Puff. His name referred to a tuft of fur that stuck out at the end of his tail. When 870 got injured in February 2013, a rival Mollies female known as Ragged

Tail took advantage of her sister's weakened condition, over-threw her, and ascended to the top position. That put Ragged Tail in the Rebel category.

Four pups were born to Ragged Tail and Puff that spring, and 870 helped care for them, despite her recent conflict with her sister. Two gray females in that litter were later collared and designated 907 and 969. One of them was destined to be a great matriarch. While 870 was helping raise her sister's litter, she appeared to be in pain and had difficulty keeping up with the pack. I got a report of a pup giving her some of its food. That was a fascinating observation, for it showed that very young wolves are capable of altruistic behavior.

When Puff disappeared in October 2013, 890, who was related to Puff and had grown up with him, took over the top male position. After 870 recovered from her injuries that fall, she reclaimed her title as the alpha female. We stopped seeing Ragged Tail around that time and we never found out what happened to her. I wondered if 870 drove her out of the pack or if Ragged Tail chose to leave after losing her alpha position.

870 then had to deal with another rival: her half sister wolf 970. For a while, 870 retained the alpha position, but she got injured again and that enabled 970 to dominate her when she was in a weakened condition. This time, instead of sticking around, 870 left the pack. Without the support of her family, her physical condition quickly deteriorated and she died. When a Wolf Project crew examined her, they listed her cause of death as malnutrition, a common fate for a wolf who does not belong to a pack.

Looking back at that time, I now wonder how pups 907 and 969 were impacted by the loss of both their parents, Puff and Ragged Tail. The two gray sisters were about six months old. That is a period when young pups still have an especially strong attachment to their mothers. They missed out on that as well as on play sessions with their father. But the entire Junction pack stepped up and raised the sisters, and this socialization helped them become fully integrated into the family. The situation reminded me of a saying from Africa: *It takes a village to raise a child.* In this case, you could say: *It takes a pack to raise a litter of pups.*

890 and 970 were now the alphas of the Junction Butte pack, but they acted more like partners running a business than a bonded pair. To add to an already complicated situation, Puff, the pack's original alpha male, came back to the Junctions in late April 2014, now collared and designated wolf 911.

911 had been away from the pack and his two daughters for about six months. It turned out he had not died, as we had assumed, but had returned temporarily to his natal pack, the Blacktails. We never figured out why he walked away from the Junctions and later rejoined the pack. No other Yellowstone alpha male had ever left a pack while being in such a dominant position and then returned.

911 seemed to be bonding with alpha female 970, and that relationship appeared to elevate him to the alpha male position once again, making 890 the number two or beta male. Although there had been intense squabbling among the females, I did not see any conflict between the two males and 890 seemed to accept his lower rank.

It was an intriguing situation. The partner of the new alpha female became the alpha male, and the consort of the deceased former alpha female got demoted. I had already seen that wolves live in a matriarchal society where the top female is the true pack leader. The stories of 911 and 890 show just how consequential it can be for a male to be allied with the right female.

IN FEBRUARY 2015, 890 did well despite his lower ranking. Since the two younger females, 907 and 969, were daughters of alpha male 911, they did not breed with their father. Both mated with beta male 890. That season, it was good to be number two. He also got into a mating tie with alpha female 970, meaning he mated with a total of three females.

The term *mating tie* refers to how the male wolf is locked internally with a female during the act of breeding. It can last as long as thirty minutes in some cases. Once a pair of wolves are in a tie, they often bed down together. Later there comes a point when the female struggles to get up, and that breaks the tie.

After 970 got in that mating tie with 890, she had a fling with a male from the Prospect Peak pack known as Twin and mated with him twice. Twin was the wolf who later joined the Lamar Canyon pack and became their alpha male. By that time, he had been collared and given the number 992. After her brief encounter with Twin, 970 returned to the Junction Butte pack, where she finally mated with 911. The Junction Butte alpha was clearly a strong-willed female wolf who did what she wanted when she wanted.

I found 970's affair with Twin especially interesting. When he arrived in the Junction territory, he repeatedly howled. 970 went directly to him and within a few minutes the two wolves were mating. I felt there was something about Twin's calls that had attracted 970. I keep track of howling throughout the year. We analyzed my data and found that during the mating season wolves howl five times more often than they do during the denning period. That may be because it is in the interest of adults to call attention to themselves when looking for a breeding opportunity, but later in the season they want to avoid advertising the location of their young pups.

That same mating season, 755 also got involved with 970's amorous adventures. He was traveling through the Junction territory before his move to Hayden Valley. 970 saw him in the distance and rushed over, making it plain that she wanted to mate with him. At that moment, 890 ran in and chased 755 away. If it had not been for that interruption, 970 would have mated with four males that year: two from her own pack and two outsiders.

970 denned on the south side of Specimen Ridge that spring. The den was near Agate Creek, about five miles from Slough Creek. The two-year-old sisters 907 and 969 had their first litters of pups in the Natal Den at Slough Creek, the same site used by the 06 Female in 2010. In total the three mothers had twelve pups, and eight survived to the end of 2015. Their fathers would have been split between 911, 890, and outsider Twin. A female pup born that spring to 969 would eventually contend with both her mother and her aunt 907 for the alpha female position.

DURING THE FEBRUARY 2016 mating season, I saw 907 reject alpha male 911's repeated attempts to mate with her. She must have instinctively balked at breeding with her own father. He accepted her refusal and walked off and bedded down. The incident reminded me of how the 06 Female often rejected mating attempts by males who displeased her. If a male was too persistent, she would beat him up. She dominated males so completely that they would not even try to fight back. Her daughter 926 was a small female but just as fierce with males. I remembered how she had dominated the four big Prospect Peak males all at the same time. I have seen 144 wolf matings over the years and I have never once seen a male wolf force a female.

Years ago, a thought came to me that might explain why male wolves accept living in a matriarchal society and do not use their greater size and strength against females. Every spring they see females go into a den and later come out with tiny wolf pups. I do not think male wolves have any understanding of how the females create those pups. Perhaps that causes the males to think that the females are more powerful than they are.

After turning 911 down, 907 went to nearby beta male 890 and flirted with him, and he mated with her right away. 907 was more distantly related to 890, so he was a much better choice to sire her pups. The two wolves mated a second time and alphas 911 and 970 also got in two mating ties. I saw 969 trying to get 890 to mate with her but missed documenting a tie. We later saw that she was pregnant, which suggests that 969 and 890 did get together.

I noticed that 911 did not do anything to 890 when he was mating with 907, despite 911's alpha male status. Before my intensive study of Yellowstone wolves, I would have assumed that an alpha male would attack and punish a lower-ranking male for trying to mate, but that was not the case here. The alpha male's acceptance of 907's rejection of him as a mate seemed to extend to not interfering with her choice to mate with another male in his place.

Over the years, I collected 101 accounts from other people who saw Yellowstone wolves mating. Adding these to my own 144 sightings during twenty-four mating seasons totals 245 breeding observations. I analyzed all those accounts and here is what I found.

The main breeders in a pack are the alpha males and alpha females (54 percent and 44 percent of all known ties, respectively). Alpha males are more likely to breed with subordinates than alpha females are. The alpha males in my sample mated with a subordinate female 15 percent of the time, while alpha females mated with a subordinate male 6 percent of the time. But when it comes to subordinate wolves, females are more likely to mate than males. The subordinate males in my sample managed to mate 20 percent of the time, and subordinate females got to mate 29 percent of the time. Intriguingly, wolves are 10 percent more likely to mate with a wolf of a different color (grays with blacks and blacks with grays) than of the same color. That would help promote genetic diversity.

People often ask me if wolves are monogamous. A study done by my Wolf Project colleague Quinn Harrison as part of his master's degree at the University of Minnesota found that 54 percent of Yellowstone alpha pairs were monogamous. I do not know what the data on people might be on that subject, but perhaps it would be fair to say that wolves are just as monogamous as humans.

THAT SPRING THERE was a fascinating sighting by long-time Yellowstone wolf watchers Becky Cox and Chloe Fessler. They told me they saw the eleven-month-old Junction pups from 2015 chase a bull elk along a ridge near Slough Creek. The bull got one of his hooves jammed in a rock pile and could not pull it out. Working together, the pups pulled him down, but did not know how to finish him off. One of the adult wolves ran in and gave the bull a killing bite. That chase, confrontation with the elk, and especially the fatal bite by the adult wolf would have served as a memorable hunting lesson for those pups.

970 once again denned on the south side of Specimen Ridge, and 907 and 969 both had their pups at the Natal Den at Slough Creek. Slough Creek was a convenient central location for the adult wolves, so they attended those mothers and their pups far more often than 970 and her litter at the more remote site on Specimen Ridge.

On April 21, I went to Slough Creek in the early morning and saw 890 bedded down near the Junction den. To the south of the den, I spotted seven wolves chasing an elk north. Three wolves in the group were collared. It took me a few moments to realize they were from the Mollies pack. Two of them were big males. Two other adults and three yearlings were in the group. All seven were chasing the elk toward 890 and the Junction den.

I looked north and saw that 890 was now with two yearlings and all three were standing and looking at the Mollies wolves to the south. Then the small group of Junction wolves ran toward the invading wolves. The seven Mollies wolves spotted them and chased the three Junctions. I saw that 890 was running off to the south, rather than back to the north. It looked like he was luring the rival wolves away from the denning females. Alpha male 911 appeared and joined the other three Junction wolves. The Mollies stopped, then went back north toward the den. 890 and 911 turned around and watched them.

I pointed my spotting scope at the den area and saw that a Junction gray yearling was now leading the Mollies away from the den in a northerly direction. On its own, the young wolf had come up with the same distraction strategy used by 890. Soon the seven Mollies wolves stopped and came together for a rally, meaning they formed a tight cluster and had a loud group howl. Then they ran back toward the den.

I checked on 911 and saw he was alone. He did a bark howl, a warning call that is a mixture of howling and barking. His calls would be heard by the mother wolves in the den to the north. The seven Mollies were now back in the den area, sniffing around. The ground there would be crisscrossed with many wolf scent trails and that seemed to be confusing them. They looked like they did not know that 907 and 969 were in the den, just one hundred feet from them.

Suddenly the seven Mollies ran off to the south in pursuit of two Junction yearlings. Like the other Junction wolves, the younger pack members led the invaders away from the den area. All those wolves went out of sight through a pass to the west.

911 continued to howl from his position to the south. Four Junction yearlings who likely heard the howls joined him. All that howling must have also warned 907, for she came out

of the den. I immediately saw that she was limping badly on her front right leg. This was a new injury and might have been inflicted on her by a Mollies wolf when I was monitoring other wolves. She held that leg off the ground as she slowly hopped downhill. Despite her injury, she did not seem stressed out by the recent presence of the Mollies in her den area.

I looked back at the den and saw 969 bedded down at the entrance with just her head sticking out. Like her sister 907, she appeared calm and confident. After looking around, 969 slipped back into the den. Within a few moments, she would be with the pups, who would be around two weeks old. 907 was still in view. I had seen eight of the eleven Junction wolves. Even without their other pack members, they outnumbered the Mollies wolves eight to seven.

The calm attitudes of the two mother wolves suggested to me that they felt safe in their den tunnel. Years earlier several females from the Slough Creek pack using that same den survived a multiday siege by a rival pack. It was a highly defensible position.

We thought the Mollies had left but I saw them coming back toward the den. A group of five Junctions to the south howled. I felt that was another warning to pack members up at the den area. 907 was now out of the den and walking off to the east. The Mollies wolves ended up on a ledge above the den. They went downhill from there to the den entrance. 907 watched them from the east. 969 was still in the den, likely just inside the narrow entrance, a position that would enable her to attack an intruder. If necessary 907 could run in, approach a wolf trying to enter the den, and bite it.

I expected the Mollies to mount an attack on the den, but they seemed to be leaving. 907, taking advantage of their retreat, rushed back to the den and slipped inside. Two Mollies stopped and headed toward the den, then hesitated. Both ended up turning back. Soon all seven invading wolves went out of sight to the west. The threat to the Junction mother wolves and their pups was now over.

I was impressed with how second-ranking 890 had taken the lead to lure the Mollies wolves away from the den in the early stages of the situation. Then alpha male 911 and the younger adults drew the rival wolves toward them and away from the den. 969 and 907 were the last line of defense regarding the pups. The potential den raid was thwarted by the courageous cooperative behavior of the Junction wolves. After the Mollies wolves left, we watched the den entrance. When the pups later came out, we got a count of nine. All were uninjured and healthy.

EARLY ON THE morning of April 30, I stopped at Slough Creek to monitor the den. I got 970's signal in the area but did not spot her. It looked like she was probably traveling north from her den on Specimen Ridge to the main Junction den at Slough. I went back to Slough in the evening and got a mortality signal from 970. It seemed to be coming from a spot west of the creek, about a mile south of the main den. Wolf Project biologists normally hike out right away to check on a dead wolf, but the creek was too high to cross. A crew finally got out there in August. They could not determine the cause of 970's death from her scattered remains but felt wolves had likely killed her.

Over the years, we found being attacked by other wolves accounts for about 50 percent of adult wolf mortality. I previously wrote that we estimate about 15 percent of wolf deaths are caused by prey animals during hunts. During the initial years after reintroduction, no hunting of wolves was allowed in the areas surrounding Yellowstone, but in 2009, the states of Montana, Idaho, and Wyoming started to introduce wolf hunting, and each year some wolves from Yellowstone packs were either shot or trapped. From 2009 to 2023, hunting was the cause of about 10 percent of wolf deaths. That means these three violent causes account for 75 percent of adult Yellowstone wolf deaths in those years.

The primary suspects in 970's death were the Mollies. They had chased Junction wolves at Slough Creek nine days before 970's death. On the day I got 970's mortality signal, twelve Mollies were just a few miles away in Lamar Valley. The previous day, April 29, I had gotten signals from a big Mollies male to the south of Slough Creek, just a mile or two from where 970's body was later found.

The Mollies had a long history of violence against packs that lived in the northern section of Yellowstone. It would be an ironic twist to the story if the Mollies killed 970, for she had been born in the Mollies pack, then had dispersed to help form the Junction pack.

The death of 970 set off a *King Lear*-type rivalry in the Junction Butte pack that would go on for years between three females: sisters 907 and 969, and 969's daughter Black Female, who would turn out to be especially ambitious.

5

Into Enemy Territory: 890's Story

THE NATAL DEN at Slough Creek was a protected space for the Junction pups, the equivalent of a fortified safe room in a family's home. It was a place where they could hide while they were learning what was dangerous and what was not. In mid-May, for example, a grizzly showed up at the Natal Den. 890 and 969 immediately chased the bear, and both bit it on the rear end. The pups did the right thing: they all disappeared into the den. A few weeks later, three huge bull bison walked into the area just downhill from the Natal Den. The pups ran into the den but soon came back out and calmly watched as the bulls grazed on grasses and continued on their way.

Not only was the den a safe place to hide from predators and enormous bull bison, it was also a space where the young pups could keep out of the elements. On a later day when heavy rain poured from the skies, I watched a black pup run into the den for shelter. Later a gray pup came out of the den, saw that it was still raining, and went back inside. The nearby yearlings and adults seemed impervious to the deluge and made no effort to seek shelter. I figured the pups would soon learn the falling water did them no harm.

The last nursing I saw was on June 12, when the pups were about eight weeks old. After that, they ate meat, mostly small pieces from elk and bison carcasses that the adults regurgitated for them. Wolves can more easily carry food internally than in their jaws, so the regurgitation system works especially well when a kill is made far from the den. Pups trigger a regurgitation from an adult by licking its muzzle. When a pet dog licks the face of a returning human friend, it is a remnant of the behavior of their wolf ancestors.

The skills and physical abilities of the pups developed quickly. By early July, they knew how to follow the scent trails of the adults around the den area. That month I watched a female yearling tease one of the pups while it was feeding on a piece of meat. The yearling hit the pup on its rear end with a front paw. The pup instantly turned and snapped its jaws at her. The older wolf continued her teasing and soon was hitting the pup with both front paws. Each time, the pup spun around and tried to bite her, but she backed up just far enough to avoid the pup's jaws. She did that twenty-five times. I was impressed by the pup's lengthy defense of its food.

On July 22, horses and mules got loose from an outfitter's camp and wandered up toward the Natal Den. I looked that way and saw 907, 890, and some yearlings keeping an eye on the approaching animals. The adult wolves did not seem too concerned. Several pups were out of the den and seemed less certain how to evaluate the danger. They apparently decided it was best to be cautious and went back underground. The horses and mules passed by without incident. By that time, the adult wolves had slipped into some nearby trees.

Three riders working for the outfitter then arrived. They rounded up the horses and mules and took them back downhill on a route that was farther away from the den. None of the adult wolves seemed particularly stressed. In the past, I had seen that wolves did not act aggressively or fearfully if riders came through their area. They just casually moved out of the way. Perhaps they classified horses as being something like elk or moose, despite the humans riding them.

One day 911 came back to the den with the head of a cow elk and gave it to the hungry pups. At that time, 911 was six years old and second-ranking male 890 was five. The two wolves had been through a lot together, ever since they had linked up with the Mollies females four years earlier to form the Junction Butte pack. Now that 870 and 970 were dead, 911 was going to have a problem in the next mating season. Since he was the father of 907 and 969, there would be no breeding with them. In contrast, despite his lower position in the hierarchy, 890 would likely mate with both.

Unlike the females in the pack, adult males 911 and 890 got along well, regardless of who was the highest-ranking. I remember a day when alpha male 911 was bedded down. Beta male 890 came over and greeted him. 911 reacted in a friendly manner by thumping his tail on the ground. 890 bedded down next to the other male. 890 was much larger than 911 and probably could have beaten him, but I never saw them fight. Perhaps 890's ability to breed 911's daughters dampened any ambition 890 might have had to be the alpha male. You could say 890 got the mating benefits of an alpha without the stress and wear and tear of being an alpha. 890 seemed to have a great life in the Junction pack. Then he did something totally unexpected.

On July 22, 2016, I spotted eight Junction pups following 890 north from the Slough Creek den. He regularly looked back to check on the pups. Soon the pups got tired and bedded down. 890 must have noticed, for he bedded down as well. A black pup, who looked like a miniature version of the big male, came over and greeted 890 by licking his face. The pups later got up and went back south, toward the den. 890 stood, walked south a short way, then stopped and watched the pups. He bedded down there and continued to monitor them. About ninety minutes later, 890 walked off to the north by himself. After losing him, I left to check on other wolves.

THAT WAS THE last day 890 spent with the Junction Butte pack. At the end of July, he was seen twenty-three miles south of Lamar, in Pelican Valley, the home of the Mollies pack.

Pelican Valley was a dangerous place for him. On January 13, 2012, when 890 was just nine months old, the Mollies had chased some Blacktail wolves and killed 890's uncle, a big male known as Medium Gray. They killed 890's mother, 830, on March 22. On June 1, the Mollies attacked his father, 838, and the wolf died of his wounds. Ten weeks later, they killed 890's cousin, wolf 777. We do not know if young 890 witnessed any of those attacks, but I think it fair to say he was well aware that the Mollies were an enemy pack.

Now, four years later, 890 was in the middle of that pack's territory. It would be like a guy wearing a Yankees T-shirt walking into Fenway Park for a Red Sox game against their hated rivals. 890 was either the bravest wolf I had ever known or the dumbest. His actions baffled me, for at the time it made no sense. I believe wolves are rational, thinking beings, so I tried to understand why he was behaving that way and why he had effectively resigned from his pack.

890 had spent the last few years with former Mollies females 870 and 970. Maybe he was seeking out a mate from that same family. We knew that the Mollies alpha male had been killed by a bull elk the previous year. Through the end of 2015 and on into the summer of 2016, the pack did not have an obvious dominant male and all the pack members were sons and daughters of the alpha female. Perhaps crucially, the three adult Mollies males who had displaced 755 from the Wapiti pack had left the Mollies pack fifteen days before 890 headed to Pelican Valley. Their departure, along with the death of the Mollies top male, created an opportunity for 890 to join the pack. He seized it and became their new alpha male. His timing was perfect. Had he arrived any earlier, he probably would have been killed.

When 890 joined the Mollies, the family consisted of alpha female 779 and her fourteen sons and daughters: nine young adults and five pups. He was now in a well-functioning pack that had a lot of females who were all unrelated to him. His risky dispersal journey had paid off spectacularly.

As I tried to figure out why 890 would go alone into the Mollies territory, I considered how a detective might investigate the case. I checked Wolf Project records and looked back over my field notes to see if I could find any clues. I saw in the Wolf Project's 2015 annual report a line mentioning that in that year the Junction wolves had left their territory in the north of the park and traveled to Pelican Valley, the home of the Mollies wolves. I was not sure if that information was from a tracking flight or based on GPS collar data. Regardless, the information suggested 890 had a certain level of knowledge regarding the Mollies wolves and where they lived.

I found a lot more relevant information in my field notes. In February 2016, we frequently saw Mollies alpha female 779 and other Mollies wolves in Lamar Valley and nearby locations. That put 779 near 890 and other Junction wolves. On the twenty-ninth of that month, I got radio-collar signals indicating that 779 and 890 were within a mile of each other. It was the mating season, the time of the year when wolves are especially interested in meeting new members of the opposite sex, even if they are from another pack. The two wolves would have been close enough to hear each other howling. If the wind was right, 890 might have gotten 779's scent, a scent that would have told him she was in season. If former Junction alpha female 970 could breed with the outsider male Twin, then the Mollies alpha female could have easily done the same thing with 890.

The Mollies wolves later returned to their home in Pelican Valley. Two months after that, 779 looked pregnant, despite having no unrelated males in her pack. She had four pups that spring. If she and 890 had met up in late February and mated, his solo march into enemy territory might well have been to find her. Later, I had a conversation with Dave Mech and he told me that hormones in male wolves are lowest in July, which was when 890 joined the Mollies. I don't know if 890 was lucky or driven by instinct, but he made his move when the males in other packs would be at their least aggressive.

I had already noticed that 890 had a great deal of self-confidence. There was also the matter of his size. When he was recollared in early 2017, 890 weighed 138 pounds, making him one of the biggest wolves we have ever had in Yellowstone. The Mollies supported themselves primarily by hunting bison, which can weigh up to two thousand pounds. Because of his huge size and great strength, 890 was an ideal recruit. The Mollies alpha female, 779, was later recollared and examined. Weighing in at 137 pounds, she was the biggest female ever known in the Yellowstone area. As the average Yellowstone wolf weighs 100 pounds, 890 and 779 were a good match.

A study by the Wolf Project staff found that the optimum number of wolves for hunting adult elk is four, but it is nine to thirteen for hunting adult bison, which are much larger and stronger. Those findings showed the importance of cooperation among pack members when hunting and how the number of wolves in the pack was something that would be particularly significant for the Mollies. The addition of a big male like 890 likely greatly improved the pack's ability to successfully hunt bison.

IN LATE OCTOBER, the Mollies came up to Lamar Valley. I got a count of eleven wolves. Alpha female 779 was in the group along with their new alpha male, 890. That October sighting was the first time I had seen 890 in the pack. Two of the family's four black pups were with the older wolves. Those black pups looked like 890, but only DNA samples could decisively tell us if he had sired them. The pack sniffed around a bison carcass recently visited by the Junction wolves. 890 would be getting the scents of his former pack-mates. I suspected he had no regrets about his decision to leave the Junctions and join forces with the Mollies wolves, for he soon departed the area with his new family.

In the fall of 2017, I drove south and hiked into Pelican Valley, the home of the Mollies. On arriving I got signals from alpha female 779 and from 890. I found the pack relaxing along a creek, near a new carcass. This was a remote section of the park without any roads. Wolves living here would see hikers passing through the valley only in the spring, summer, and fall. I think that if I were a wolf, I would rather live here than at Slough Creek, a location with a lot of people and cars.

A few weeks later, I spotted 890 and the Mollies wolves in Lamar just south of what had originally been the Druid den forest, a location that was now being used by the Lamar Canyon pack. I saw the Mollies kill a bull elk. Lamar alpha female 926 and the rest of the Lamar wolves were visible to the north. I could see 926 looking directly at her family's longtime enemies, the Mollies. The Lamar wolves howled defiantly at the invaders and the rival wolves howled back. But the Mollies wolves stayed where they were and did not bother the Lamar family.

In late December, I was watching the eight Junction wolves at Slough Creek. I got signals from 890 to the north of them, then heard the Mollies pack howling from that direction. The Junctions repeatedly howled back at the intruders and the Mollies howled at them. Then the Junction wolves started to march toward the other pack. When I next heard the Mollies howling, it sounded like they were moving away. The Junctions were soon where the other pack had been howling, but by that time the Mollies were gone. That was a second case where the Mollies wolves could have charged at a pack much smaller than theirs and likely killed some of them. But they chose not to be aggressive and moved off. Later I saw the Junctions sniffing at a site where the Mollies had been. That would mean they got the scent of their former pack member and would know for sure he had joined the rival pack.

Before 890's arrival, the Mollies had been known as the most violent pack in Yellowstone. As far as I know, during 890's tenure as alpha male they never killed any wolves. I felt that showed that one wolf can change the longtime culture of a pack—in the Mollies' case, a tradition of belligerent behavior. 890 seemed to place a higher value on being accepted into a pack with plenty of unrelated females than on continuing a long-running feud, even though that feud had claimed the lives of some of his own family members.

Wolf 890 was a good example of a successful Disperser. As a young wolf, he left the Blacktails, his birth pack, to cofound the Junction Butte pack. Years later, he walked away from the Junction pack and soon became the alpha male in the formidable Mollies pack. 890 sired pups in the Mollies pack for as many as five years, meaning he has descendants in that central section of Yellowstone as well as in the northern portion where the Junctions are still based.

MY WOLF PROJECT colleague Quinn Harrison examined records of 123 dispersing park wolves in the 1995–2016 period. He discovered that most dispersals take place between the fall months and the February mating season, and that males are more likely to disperse than females (57 percent versus 43 percent). That timing coincides with a rise in male hormones and would seem to explain an urge to leave their family to seek out an unrelated mate. 890 was unusual in this respect, as he dispersed in July. But his story is different from most dispersing wolves, for I think he got the Mollies alpha female pregnant the previous February. Perhaps he simply wanted to renew his relationship with her.

Dispersal is a risky strategy. A 2010 research paper with Doug Smith as the lead author reported that Dispersers are twice more likely to die than wolves who stay in their natal pack. However, Dispersers who survive and join a pack or start a new pack have a much higher chance of being breeding wolves than their packmates who stay home. Dispersal is an especially effective strategy for yearlings. Quinn found that among the wolves he studied, a male yearling has a 6.7 percent chance of breeding if he stays with his pack and a 28 percent chance if he disperses. Female yearlings who stay with their pack have a 33 percent chance of having pups, compared with a 42 percent chance if they disperse. His data on female yearlings is similar to my observation that 29 percent of the matings I have seen within a pack involved subordinate females, meaning non-alphas.

Another finding in Quinn's study was that if a wolf is going to take the risky step of leaving its family, the most successful strategy is for the wolf to start its own pack.

Dispersing wolves that form new packs are eleven times more likely to reproduce than Dispersers who join existing packs. Once again, 890 proved himself to be an exceptional wolf. Despite being old, he successfully joined an existing pack that had plenty of females who were unrelated to him and took over its vacant alpha male position.

I LAST SAW 890 in March 2021. He was with a group of Mollies wolves near Lamar Valley. 890 seemed to be in good health and had maintained his position as the Mollies alpha male for nearly five years. By then, he was ten years old, triple the average life span of a Yellowstone wolf. He had outlived 779, his first mate in the Mollies, and now was with a much younger female. During 890's tenure as alpha, the Mollies maintained their less aggressive stance toward other packs.

The last known sighting of 890 was a photo taken by a trail camera that summer. His once glossy black coat had turned gray and it looked like he had lost a lot of weight. Then, like 755, another black wolf who had turned gray toward the end of his life, 890 faded away into the wilderness.

6

Won't Back Down: 911's Story

AFTER 890 LEFT the Junction pack in July 2016, alpha male 911 carried on without his cousin. He was six years old but all the injuries he had suffered—likely during hunts to feed his family—made him seem much older. 911 often fought elk three to seven times his weight. One day I saw that he was limping on two legs and had a wound on his shoulder. I thought of the many NFL players who suffered in their later years from all the hits they had taken during their careers, hits similar to a kick from an elk hoof.

On the morning of September 14, about seven weeks after 890's departure from the Junction pack, I spotted eight wolves from the Prospect Peak pack in Lamar Valley. They were south of the Lamar River, near the outlet of Amethyst Creek. The group included alpha female 821, alpha male 966, and three other adult males. The Prospects were the wolves who had killed Lamar alpha male 925 the previous year, so were known as an aggressive pack. I then saw a cow elk standing in a section of the Lamar River, north of the wolves. One wolf waded out into the river toward the elk, then turned back when he reached deeper water.

Soon the eight wolves walked off and went up a hill southeast of the elk. They bedded down in a spot that gave them a good view of the river. I looked at the cow again and saw that she was injured. The Prospect wolves must have attacked her during the night. She had apparently fled to this section of the river and was now standing her ground in about four feet of water.

The elk was tall enough that she had a solid footing on the riverbed, but wolves would have to swim out to her, putting them at a huge disadvantage if they tried to attack her. An elk standing in deep water is in a strong defensive position, as it can rear up and strike down with its front hooves at a swimming wolf. This cow likely had dealt with wolves many times before and would know the river would be her best defense.

I put my high-powered spotting scope on the elk and saw a wound at the base of her tail. Blood was seeping out and dripping down one of her hind legs. That injury and the cold water would gradually weaken her, giving the wolves a better chance to finish her off.

Prospect alpha male 966 and another wolf got up and stalked toward the cow. The big male turned back, probably because he saw how defiant she was. The other wolf waded out through a shallow section of the river. The cow held her head high as she stared at the approaching wolf, then she charged. He darted back and forth as she carried the fight to him.

Soon the two were in deeper water and the wolf was forced to swim as he continued to confront her. He dog-paddled toward her, then reared up and bit into the front of her throat. The bite would be fatal if he could hold on long enough, but he did not have enough leverage and had to let go.

By that time, the alpha male had swum out and positioned himself behind the elk. He bit into the wound on her rear end. As she swung around to confront him, he lost his grip. When she went on the attack, he gave up and swam back to the south bank of the river. At that point, alpha female 821 arrived. She stared at the elk and seemed to be evaluating her condition. Apparently deciding the cow was still too strong, she bedded down to watch her.

A short time later, all the Prospect wolves moved off and scavenged on an old bison carcass. After getting what scraps they could, the pack disappeared into a forest. I figured they would rest there during the hotter hours of the day, then make another attempt on the elk in the evening. I glanced at my watch. It was 10:07 a.m.

I checked for signals from the Junction wolves and got alpha male 911 and female 969. I then heard that a black Junction wolf had been seen chasing an elk north of the road, across from where the cow elk was standing in the river. I scanned the area and saw the black traveling north. Then I spotted three other Junction wolves in the rolling hills behind the black, including 969. She and her sister 907 likely had their pups with them up in that area. I took a break at that point and went back to my cabin. It was 1:34 p.m.

I got back to that area around five in the afternoon and heard from wolf watchers that a collared gray wolf had come in from the west and bedded down close to the cow elk on the north side of the river. I set up my spotting scope and saw that the cow was standing in the same spot as when I left. Then I looked at the gray wolf and saw that it was 911. He had reportedly made an attempt on the cow elk but had backed off and climbed out of the river when he saw that she was aggressively defending herself.

I looked farther south and saw the eight Prospect wolves bedded down on a slope above the river. They were all asleep. I got a report that two of them, both blacks, had earlier gone downhill toward the elk and 911. At the time, he was still bedded down on the north bank of the river. The two blacks swam out to the elk but did not challenge her. They came out on the north side, approached to within a few feet of 911, then swam back without bothering him.

I pictured 911 jumping up and growling at the Prospect wolves with his tail raised. But when I asked people what 911 had done to scare them off, they told me he had just lifted his head and stared at the blacks. One of those two wolves was much larger than 911 and in good health. 911 was in poor condition and had lost a lot of weight, yet he intimidated the two males just by staring them down.

I then got more information on 911. At one point, he had waded out into the river toward the elk. She reared up and struck out at him with her front hooves. He dodged those blows and made another attempt to approach her, but she chased him off. 911 rested for a while, then made a third attempt on her. Once more he was driven off by her aggressive counterattack.

When 911 got up, I noticed that not only did he have a limp in his left hind leg and left front leg, but his hind right leg was now injured as well. Despite his injuries, he seemed determined to finish off that cow elk in the river. He lay back down and watched her intently, probably evaluating her condition. Then he got up and limped on three legs toward the water.

His hind legs looked especially bad. They were bent more than normal and appeared to be barely supporting his weight. I also noticed his back was arched in a way that suggested walking was causing him a lot of pain. I guessed that when he had approached the cow elk she had reared up and come down on his back with her front hooves. I took another look at the cow and saw that she was a big animal, probably close to five hundred pounds. 911 had been losing weight and looked to be way under a hundred pounds. That meant his opponent was more than five times his weight.

911 had spent his adult life hunting elk to support his family. Now that he was old and much slower because of all his injuries, he needed to cut back on active hunting and let younger wolves in his family take over that responsibility. But I had come to know 911 well and guessed that nothing would stop him from continuing to hunt. That was who he was: a wolf who hunted elk to feed his family.

911 waded into the river at 6:24 p.m. The cow saw him approaching and turned to face him. He swam straight toward her, positioned himself in front of the cow, then lunged out of the water at her throat. She reared up and kicked down at him with her front hooves. The wolf dodged her hooves, swam forward, and grabbed her shoulder in his jaws. She broke free and tried to get away but 911 followed and bit into her rear end. She rushed through the water, then circled around. That change in direction caused him to lose his grip.

Looking exhausted, 911 swam off and climbed out of the river. That was his fourth failed attempt to take down the big elk. Now his hind legs seemed even worse. The wolf looked like he had no strength to continue.

Despite the terrible pain he must have been enduring, 911 went right back in the cold water. In his weakened state, he must have been relying on willpower alone. He positioned himself to make another attempt to grab her throat. In response she once again reared up and kicked down at him. The wolf managed to dodge the blow. That was 911's fifth attempt on her. I noticed that the cow's head was drooping down, a sign her strength was ebbing away.

Her injuries and time in the cold water had taken a heavy toll and she seemed near the point of collapsing. Time was on the wolf's side. I felt the smart thing for him would be to just wait her out from the riverbank. 911 came out of the river. But instead of resting, he went right back in for a sixth try to get her. The cow waded toward the north bank and that partially blocked my view. I got glimpses of the wolf going after her and the cow trying to fight him off.

I moved to a new position and from that spot saw that the epic battle was over. 911 was standing on the riverbank and had already begun to feed on the elk. It was 6:34 p.m. His last three attempts on the cow elk had spanned just ten minutes. It was the sixth bout with her that finished the fight. Now that I had time to think about what I had witnessed, I realized that 911's fight with the elk was like Muhammad Ali's 1974 championship bout with George Foreman in Africa. That was the fight where Foreman pounded relentlessly on Ali for seven rounds. After taking all those blows, Ali saw that his opponent had tired himself out. He went after Foreman and knocked him out.

A woman let me look at photos that she had taken of the final seconds of the fight. It showed 911 jumping out of the deep water with open jaws, about to grab the cow's throat, the classic finishing move of an experienced wolf. The most startling thing about the photo was seeing how small the wolf was compared with his opponent.

So much had gone on in the last few minutes that I had forgotten about the Prospect wolves. They had wounded the elk earlier in the day, but despite having eight combatants, the pack had failed to take her down. 911 did that by himself, even with three injured legs. All eight Prospects were now staring at 911. That was not a good sign.

The situation then got worse. Prospect alpha female 821 and their big alpha male, wolf 966, were coming downhill and heading toward the kill site. Soon there were five Prospect wolves across the river from 911. As they got closer, 911 stood up on the north bank. He was unsteady on his three injured legs and acted like it hurt to put any weight on them. At that moment of extreme danger, he did something astonishing: 911 bedded down and calmly watched the other wolves. That action seemed to convey his confidence in being able to handle the situation.

I now saw that 966 and another wolf were wading into the river and going toward 911. But 966 turned back partway out. The other wolf swam over to the carcass, then came out on the north bank, a few lengths from the still-bedded 911. I studied the old wolf and he seemed totally calm. In contrast, his rival appeared to be losing his nerve. He tucked his tail between his legs and dropped into a defensive crouch, a sign he was intimidated by 911.

Then 911 tried to get up, but with three injured legs, it took him a long time. When he was finally standing, he walked toward the rival wolf, his teeth bared. That was too much for the Prospect wolf and he walked off. I looked at 911 again and saw that he was hunched up as though he was in terrible pain.

Then the Prospect alpha pair, 966 and 821, and another big male wolf ran to the south bank of the river and waded into the water. Soon 966 reached the north bank and went right toward 911 with his tail raised. It was now two alpha males confronting each other. 966 looked like he was in prime condition, while 911 was barely able to walk. 911 limped away from the much bigger male but did not seem scared.

It looked like 911 had realized the eight wolves were too many for him to handle. He started to head toward the road to the north, where the rest of his family was. If he could get to them, he would be safe. That would mean abandoning his kill to an enemy, but it would save his life.

I turned my scope back to the other wolves and saw that they seemed hesitant to approach the fresh carcass. One wolf was wading back through the river to the south side. I guessed that the Prospect wolves might suspect that this situation could be an ambush and that all the Junction wolves might suddenly appear and attack them.

I checked on 911 and was shocked to see that he had turned around and was going back toward the river. He was looking at the other wolves, seemingly unconcerned about their presence near his kill. Two of the Prospect wolves spotted him and trotted his way, but soon turned back. 911 had bedded down by that time and was calmly looking around. I saw 966 go to a spot where 911 had rested earlier, and the big male sniffed the site. 911 raised his head to get a better look at the rival alpha male.

At 7:25 p.m., 911 got up and moved toward the carcass. After a short distance, he stopped and sat. I got the sense that the pain in his three legs had gotten too much for him to bear and he had to rest. He watched the other wolves, then bedded down with his head on the ground. It was an astonishing sight, for he appeared completely unconcerned about the close proximity of the eight Prospect wolves.

A gray wolf from the other pack came toward 911 and stopped one wolf length from him. 911 calmly stayed bedded down, giving the impression that he had no fear of the other wolf. He did not even bother to look up at the gray. When he did turn his head toward the wolf, it immediately backed off.

I had seen what terrible physical shape 911 was in and could only guess how much pain he was enduring. The thought came to me that 911 understood how important it is to project confidence when you are facing an imminent threat. It was a bluff that was working.

But soon the alpha female and a black wolf joined the gray near 911. That seemed to give the gray more courage. At that point, 911 stood up. Perhaps because they finally saw how injured he was, the three wolves charged forward and attacked 911. They threw him to the ground and all three bit him. Three more Prospect wolves ran in and joined the attack.

911 got up and valiantly fought with his enemies. Aston-
ishingly, rather than running away, I saw him charge at other
wolves and bite them. That moment forever defined 911's
character for me: despite being in imminent danger of violent death from a far superior force, he chose to move toward
the threat. Two more Prospect wolves ran in, making it eight
against one.

Now I could just barely see 911 among the eight wolves
that were surrounding and attacking him. At 7:38 p.m., the
fight seemed to be over, for some of the Prospect wolves
began to walk away. It was sixty-four minutes after 911 had
spent all his strength killing the elk. As the other wolves
departed, I could see his battered and bloody body. 911 was
dead, but he died in the most heroic manner imaginable. It
was a fitting way to complete the life story of the last founding member of the Junction Butte pack.

BY THEN IT was getting dark and there was nothing to do
but head home. On the way, I tried to figure out what had
been going on in 911's mind. I was especially mystified about
why he did not save himself by retreating north, toward the
park road. If he'd crossed it, he could have joined his pack
a mile or so farther away and would have been safe. Then I
realized something. The Prospect wolves, after feeding on
the elk, could have easily picked up 911's scent trail and followed it across the road during the night. That trail would
take them right to the Junction wolves. If he had gone up
there, 911 might have saved himself, but he also would have
been putting his pack in grave danger.

When he fought to the death with the rival wolves, 911
was positioned between them and his family. That included

his adult daughters 907 and 969 and their pups, who would be 911's grandsons and granddaughters. An alpha male's most important responsibility is to protect his family. 911's decision to stand his ground and put himself in harm's way, despite facing odds of eight against one, was the most courageous thing I had ever seen in my life.

I began writing this section around the time of the twentieth anniversary of 9/11. *60 Minutes* aired a segment then about the heroism of the first responders, especially the New York City firefighters, who rushed to the Twin Towers after the two planes crashed into them. Their mission was to bring the seventeen thousand people in the burning upper floors down the stairs to safety. Like the wolf known as 911, they willingly faced off against a terrible threat. That day, 343 New York City firefighters lost their lives when the towers collapsed. I would also like to pay tribute to the passengers on Flight 93. Like 911, they did not back down. In their case, they charged forward to confront the terrorists on their plane.

My first National Park Service job was as a firefighter. I was on a helitack crew in Sequoia National Park. Our helicopter pilot would fly us out into the Sierra Nevada mountains where lightning had started a fire and drop us off. We camped out there until our fire was out and ate Vietnam-era C rations for all our meals. Prior to that, when I was with the US Forest Service, I worked a hundred-thousand-acre fire in Washington state. Because of those jobs, I felt a special connection to the 343 fellow firefighters.

When the wolf we knew as 911 was captured and radio-collared in late 2013, he was assigned the next number in sequence. At the time, we did not consider that the 911 designation had any special meaning. After watching the way he lived out his final day of life, I will always link his hero-ism with the extraordinarily brave men and women who lost their lives on 9/11, 2001.

If my readers might ever wonder what it would be like to be a wolf, think of 911. He was a classic alpha male who wholeheartedly accepted the responsibilities of the position. He worked hard to feed and protect his family, up through the last day of his life. 911 treated the other adult males in his pack well, including his cousin 890. He even showed tremendous patience when the females in his family were squabbling.

When a Yellowstone wolf like 911 dies, biologists from the park's Wolf Project come out and examine its remains. The crew weighed 911 and found that he was only sixty-seven pounds. He had lost a third of his normal weight because of recent injuries. I had noticed his three injured legs, but there was something far more serious about 911's condition. His lower-left jaw had been broken several months earlier and had never healed. Like the injuries to his legs, that injury was almost certainly caused by an elk kick. The gap created by the blow was close to a half inch, too wide for the bone to heal on its own. The broken jaw explained his great weight loss. The pain when he tried to eat would have been un-imaginable. Yet he had lived for several months after the break. That discovery of his broken jaw made 911 even more courageous than we thought. How could a wolf with a broken jaw stand his ground and fight eight enemies when his teeth were his only weapon and that weapon was now useless?

To put that in different terms, when 911 stood his ground and fought those other wolves, it was not a situation where his chances of winning the battle were slim—they were zero.

In the *Yellowstone Wolf Project Annual Report* for 2016, there is a page dedicated to 911 entitled "The Hard Life of a Yellowstone Wolf." It includes a photo of 911's cleaned skull that shows the gap in the lower jaw. Here is how the report described that broken jaw:

> The severe injury was several months old with extensive calcification, the body's long-term attempt to heal trauma to the bone. But the jaw of a wolf is constantly moving and it could never heal correctly; instead forming a gap all the way through the mandible with bone shards scattered throughout. A deeply painful injury that would have put any human in intensive care and consuming fluids for months, this wolf was still living and hunting and traveling miles in the Yellowstone backcountry, until the very last day.

7

The Rivalry Continues: 907 and 969's Story

WHEN 970 DIED in spring 2016, 907 became the next Junction Butte alpha female. Since the formation of the pack in 2012, there had been three alpha females: former Mollies wolves 870, Ragged Tail, and 970. 907 was therefore the fourth alpha female in as many years and the first one born into the Junction pack. In contrast to the rapid turnover of alpha females, there had been only two Junction Butte alpha males in those four years: 890 and 911. However, by mid-September 2016 both males were gone. 890 had dispersed in late July and 911 had died after being attacked by the eight Prospect wolves on September 14. That left fourteen wolves in the family: seven adults and seven pups.

On September 26, we heard that three wolves had been shot in Montana hunting unit 316. That area is only five miles north of the Slough Creek den, so they were probably Junctions. On September 29, we saw the Junctions and got a count of eleven. Two yearlings and a black pup were missing and never seen again. They likely were the wolves killed in the hunt.

By the end of 2016, the Junctions had only seven members: the two adult females (alpha 907 and her sister 969), a young gray female, another young wolf known as Black Female, and three surviving pups (one gray and two blacks born the previous spring). Several young adults and a second uncollared black female were unaccounted for.

The death of 911 in September 2016 had left the pack with no adult males and the breeding season was only a few weeks away. Sisters 907 and 969 needed to quickly recruit some new males into their family. The Junction pack had a high-quality territory, so any dispersing male would hit the jackpot if he wandered into the area.

IN MID-FEBRUARY 2017, I found the Junction wolves at Slough Creek and saw that several outsider males were with them. We identified them as wolves from the Prospect Peak pack: 996, 1047, 1048, and an uncollared black who became known as Black Male. They appeared to have dispersed as a group. 1047 was three, the oldest in the group. 969 was not around at the time, so it was up to alpha female 907 to interact with the potential recruits to the family.

A Junction black pup did a play bow to 1048 and the Prospect male accepted the invitation to play. The big wolf's positive response to the pup likely reassured 907 that the newcomers were friendly. Two days later, 1048, who was just twenty-two months old, mated with 907. The mating tie went on for some time and the young male fell asleep part-way through. I would later see that when 1048 got older he became much more attentive to his female friends.

When 969 returned, she seemed taken with male 1047 and did play bows to him. By then, it was clear he was the highest-ranking male in the group. 1047 had been born into the 8 Mile pack. When he was a pup, he left with other pack members to join relatives who had just formed the Prospect Peak pack. If he joined the Junctions, this would be his second dispersal.

Did 907 and 969 know that Prospect Peak wolves had killed their father five months earlier? I thought a lot about that and concluded they probably did not, for 911 was alone that day.

I had gotten signals from Prospect male 996 on the day of the attack on 911, so he was one of the eight wolves there. At the time of the fight, 1047 and 1048 had not yet been collared. In my field notes for that day, I noted that there were two uncollared black yearlings in the Prospect group. 1047 would have been older than that, so he was not involved in the attack. Later, after watching 1048 for many years, I saw that he did not have an aggressive personality. I never witnessed him bully another wolf. That suggested it was unlikely he was one of the attacking wolves. That left Black Male, and I could not determine if he was there that day or not.

The four Prospect males were still with the Junction females in March, meaning they had been fully incorporated into their pack. The new genes and experiences the four males brought with them were just what 907 needed to revitalize her struggling pack.

Late that month, sixteen wolves from the Mollies pack paid another visit to Lamar Valley. I saw 890 and alpha female 779 in the group. They traveled west through the valley and I eventually lost sight of them. They had not come near to any of the Junction wolves and stayed away from the Lamar Canyon wolves.

About six weeks after the large group of Mollies wolves passed through Lamar Valley, a subgroup of five young pack members returned. I spotted five Junction wolves heading their way: the four adult Prospect males and 969. They saw the intruders and chased them. A black male yearling in the Mollies group lagged behind his packmates. The Junctions caught him and beat the young wolf up. They could have killed him but left him and went after another Mollies wolf. The yearling jumped up and ran off to the north. He swam the Lamar River and came out on the far side. From there he continued north and bedded down just a short distance south of me, close to the park road. I saw numerous bite marks on his body.

A big crowd soon gathered on the road close to the injured wolf, and people started walking out to take photos. Knowing that would add stress to his already difficult situation, I put up closure signs to keep people away. He recovered well enough after four days to start traveling, and I last saw him heading back to his family's territory.

907 LOOKED PREGNANT that spring. She denned at a remote location, miles south of the den at Slough Creek. I wondered if she had gone on a hunt late in her term and suddenly had to give birth where she was. 969 did not seem

to be pregnant that spring. That meant the Junction wolves were not using the Slough Creek den that year. 969 traveled full-time with the Prospect males and younger Junction wolves. She acted like the pack's alpha female in that group.

Then a strange event took place. A female relative of the Prospect males showed up at Slough and chose an alternate site known as the Sage Den to have her pups. It was located downhill from the Natal Den, the original den. The Sage Den had been dug by a badger and later was enlarged by the Junction wolves. The Junctions often passed through that area and even bedded down nearby at times, but they never bothered that female or her pups. This was another case where the pack chose to spare the life of a wolf that they could have killed, even an outsider that was using one of their dens.

A second female showed up at the site and played with the mother wolf. She must have been another Prospect wolf. Later more Prospect wolves visited. By early May, we were seeing five pups at the den. Both the mother wolf and the second female nursed the pups. That implied the other female had lost her own pups but was still producing milk.

None of 907's pups survived that spring at her isolated den. I had been concerned that the pack was not giving her much support at that remote location. I wondered if her sister 969, a rival of 907's, was influencing the other wolves regarding their inattention to 907 and her pups. When 969 retained the alpha position after 907 rejoined the pack in late May, I felt it was partly due to the power of incumbency. 969 had gotten used to being the top female and the other pack members were used to having her in charge. Since the change in leadership did not involve a physical fight, I classified 969 as a Bider. For 907, the combination of losing her pups and the loss of her alpha position to her sister likely meant that 2017 would be the worst year of her life.

969 was now the fifth Junction alpha female. While the females in the family seemed to be constantly vying for supremacy, the four new males in the pack got along well with one another. 1047, the oldest of the males, took over the alpha male position without any obvious objections. He was the pack's third alpha male and he would retain that position for many years.

ONE DAY IN late May, a nine-hundred-pound cow bison came into the Slough den area. A scared Prospect pup, who was well under ten pounds, raced over to its mother. She picked it up and ran off, but the pup fell out of her mouth. The big bison went right up to the pup and sniffed its face. In response the little wolf defiantly snapped at the cow and she flinched. Then the pup casually walked off, seemingly unafraid of the massive animal. The mother wolf tried to get her pups into the den but they wanted to play. Eventually they got hungry and came to her for a nursing. After the feeding, she lured them all into the den.

Seeing that cow bison reminded me of the time I got involved with a bison rescue operation. It happened one fall when I was driving west through the park to check on the 8 Mile pack. My friend Stacy Allen was with me that day. He is the chief law enforcement ranger at Shiloh National Military Park in Tennessee, and likes to spend his vacation time in Yellowstone helping us study the wolves.

As I approached the area known as Blacktail Ponds, I saw a large crowd gathered. Two Yellowstone law enforcement rangers were on the scene. Something was obviously happening, so I pulled over. Stacy and I went to the rangers to see if we could help. They told us that a bison calf had fallen into one of the ponds and could not get out. The calf was frantically trying to climb out on the western side of the pond, but the bank was too steep for the young animal. Over the years, many other bison had drowned in the pond.

The two rangers had already decided to pull the calf out. One of them had grown up on a ranch and had a lot of experience lassoing horses. He left us and drove to Park Headquarters to get a rope. When he returned, the four of us walked out to the west side of the pond where the calf was still pawing at the bank with its front hooves. We could see that it was getting exhausted. The ranger made a lasso and got it around the calf's neck on the first try. Then we all grabbed the rope and pulled it toward us. The body of the calf hit the vertical bank and got stuck there. No matter how hard we pulled, the mass of the calf prevented us from lifting it up and over the bank.

Then Stacy, who came from a farm family, saved the day. He told us to hold on to the rope while he ran to the edge of the pond. Stacy crouched down, reached out, and grabbed each of the calf's small horns. He deftly turned the calf so that it was floating on its side. Then he rushed back and told us to pull as hard as we could. We all did that and the calf slid easily out of the water and onto the top of the bank. The big crowd that was watching us let out a cheer. As soon as she saw her calf on land, the mother bison rushed toward it. We got back to the road in time to see mother and calf trotting off together.

I had previously been involved in another wildlife incident where rangers had to save a suffering animal. Back in the 1990s, I spotted two bull elk who had their antlers entangled in what looked like a net made of thick rope. It probably was a cargo net left over from the 1988 Yellowstone fires. They had apparently been in a big fight, for one bull was lying motionless on the ground. The other bull could not break out of the netting that attached him to the dead one. I called the park dispatch office and two rangers were sent out. We hiked out to the area and one of the rangers got a tranquilizer dart in the live elk. When the elk lost consciousness, they cut the netting off him. The rangers left and I monitored that bull until he recovered from the drugs and walked off.

On another occasion, a Wapiti pup fell into a geothermal pool known as a mud pot in Hayden Valley and could not get out. A group of rangers pulled it out using a tow strap. They freed the wolf and it walked off, seemingly okay, except for being covered with mud.

ONE MORNING IN June, I found the Junction pack in Lamar Valley. The wolves were in a playful mood. Three of the Prospect males got in a chasing game where they romped around like young pups. A young female came over and two of the big males chased her. Then 907 and 1048 played together. Soon after that, I saw an example of how the male-female dynamics in the pack worked.

I was watching sisters 907 and 969 feeding together on a new carcass in Lamar Valley without any aggression or problems. The four big males were off to one side, looking like they were waiting their turn. Two of the males moved closer but stopped before reaching the carcass. Then second-ranking female 907 walked off. One of the males made the mistake of approaching alpha female 969 while she was still eating. She lunged at him with open jaws and he immediately backed off. For a while, both males just stood there wagging their tails and waiting their turn. Soon the pair cautiously moved forward. 969 looked up and threatened one of the males but let the other one eat next to her.

That incident was a clear example of the power of alpha female wolves. 969 decided who could eat and who could not. She later walked off and only then did that second male feel it was safe to go to the carcass and feed. What made that incident so significant was the history behind it. 969 was a daughter of 911, the wolf that some of those Prospect males had killed. Now they were in her pack and she was clearly the boss of them.

Another incident around that time showed who was who in the female hierarchy. Beta female 907 pinned Black Female, who was third in rank. Then alpha female 969 came over and pinned 907. After that, 907 went to alpha male 1047 and played with him. 969 saw that and apparently did not like it. She came over and got between them. The incident reminded me of the time years earlier when young 926 had gotten between her older sister and male 925.

That month, I saw that 1047 had developed a limp. He often held a hind leg off the ground when he had to trot at a fast pace to keep up with the pack. The leg never fully healed and he limped on it for the rest of his life. I did not know what caused that injury but suspected it was a kick from an elk. Later I saw 1047 limping on two legs. Luckily, the second injury turned out to be just a temporary inconvenience and he was soon back to just one bad leg.

I noticed a subtle thing that was going on among the four males. Alpha 1047 regularly exerted his dominance by jumping on the backs of fellow males 996 and 1048 and pinning them. That behavior often looked more like a ceremony than an aggressive act. I saw that Black Male tended to avoid 1047 when he was reinforcing his alpha position, and when 1047 did get on his back, Black Male did not go down. That made me think Black Male, who was the youngest of the four males, might challenge 1047 for leadership of the pack.

8

The Nanny: 1109's Story

AN UNCOLLARED YOUNG black female in the Junction Butte pack was unaccounted for at the end of 2016. We assumed she had died or dispersed. But in the fall of 2017, she reappeared and rejoined the pack. Later that year, she was captured and collared and assigned the number 1109. We think 911 was her father and 970 was her mother. If this were a Disney movie, they would call 1109 a princess, but when both her parents died in 2016, other wolves took over the alpha positions and she was just another low-ranking wolf in the pack.

Perhaps because she lost her mother and father when she was young, 1109 developed an independent personality. She definitely fit my definition of a Maverick. Maverick wolves do what they want whenever they want. They frequently break the conventions of wolf society, such as the need to

belong to a pack for protection from rival wolves and for better hunting success. When 1109 was with the Junctions, I
had the impression that she did not care much about her
position in the hierarchy. She was usually the lowest-ranking
adult female and was frequently pinned and harassed by
higher-ranking ones, including young Black Female, who
was her half sister. 1109 often left the family and lived as a
lone wolf, then returned when she felt like it. I respected her
independent spirit.

When I thought more about 1109's seeming indiffer-
ence to advancing in the hierarchy, I realized something:
both of us were Mavericks who did not care about status. I
worked for the US government nearly all my adult life but
was happy staying at a low level because jobs in the field are
more rewarding and enjoyable than higher-paying office jobs.
My pay grade for my first US Forest Service job when I was
eighteen was GS-3. Decades later, when I finished up with
the National Park Service in Yellowstone, I had only risen
to the lowly GS-6 level. That suited me, for I never cared
about climbing the bureaucratic hierarchy. I am proud to
say I never had a government office, a government desk, a
government phone, or a government computer, and never
wanted any of them.

At the end of 2017, the Junction pack numbered eight
adults: four females and four males. The alphas were still 969
and 1047. The beta female was 907 and the second-ranking
male was 1048. Two other former Prospect wolves, 996 and
Black Male, along with Black Female and 1109, rounded
out the pack. There were no surviving pups that year. That
was also the story in the Mollies pack and in 926's family in

Lamar Valley. Once again, the pup deaths were probably due to distemper. For the mother wolves in those families, there was nothing they could have done to save their pups from that plague.

I RETIRED FROM the National Park Service in February 2018 and Jeremy SunderRaj took my place with the Wolf Project. He had graduated from the University of Montana a few years earlier with a degree in wildlife biology. Jeremy started coming to the park when he was ten years old and saw his first wolves here when he was eleven. During his high school years, he had a summer job in my town of Silver Gate. Jeremy went out early every morning to watch wolves, came back and worked his shift, then went back out in the evening.

After I retired, I kept going out into the park every day to study wolves on my own time, and Jeremy kept me filled in on what the Wolf Project was documenting. At that time, the park estimated that our northern section of Yellowstone had thirty wolves, along with thirty grizzlies, thirty mountain lions and 144 black bears. For the entire park, an area of 2.2 million acres, there were fifty-two adult wolves and forty-four pups, a number close to the longtime average of one hundred wolves.

As breeding season approached, the Junction wolves visited their den site at Slough Creek. Some females went into the Natal Den, likely checking it out to see if they wanted to use it this year. One day I saw alpha pair 969 and 1047 sitting side by side. 969 stood up and averted her tail to 1047, a sign she was getting ready to breed. Black Female also flirted with

1047. Both she and 1109 would be old enough to have pups that spring.

Third-ranking Black Female seemed to have ambitions to attain a higher position in the pack's female hierarchy. I saw her pin the much older beta female 907. Alpha female 969 put the younger wolf in her place by pinning her. The rivalry between 907 and Black Female would go on for another four and a half years.

In February a film crew from *60 Minutes* came to the park to do a story on our wolves. Bill Whitaker was the correspondent, and I showed him the Junction pack through my scope. Given his many years as a television correspondent, I am sure Bill has seen a lot of spectacular things throughout the world, but he was as excited as a young kid to see the pack traveling along the side of a ridge.

That month we saw alphas 969 and 1047 mate four times. In March we spotted 969 coming out of the Natal Den at Slough Creek. She was likely getting it ready for her pups. Later I saw 969 go back into the den and clean it out. 1047 also entered the den tunnel and totally disappeared. He was a big male, so that was an indication of how tall the entrance was.

Meanwhile, 1109 was often away from the other Junction wolves, apparently content to be on her own. She did return as we got closer to the birth of the pups, but Black Female often beat her up to make sure 1109 knew who was dominant. The males wisely stayed out of all the fighting for rank among the females. Over my many years of watching wolves, I have never seen a male wolf intervene when two females are squabbling. I figured that was a sign of their intelligence.

IN MARCH I saw one of the young Junction females rolling on the snow, probably trying to scratch an itch. But she did that on a slope and started to slide downhill. A small cliff was directly below her. The wolf slid over the cliff, then fell over eight feet into soft snow that cushioned her landing.

Wolves have thick fur that fully insulates them from the coldest temperatures the Yellowstone area experiences. There have been countless times over the twenty-four winters I have spent here when I have been shivering despite wearing ten or eleven layers of clothing. On most of those occasions, I was watching wolves traveling or sleeping and they never seemed to be affected by the cold, even when it was minus 50 degrees Fahrenheit (–45 Celsius).

ON APRIL 1, in an area about ten miles west of Slough Creek, I watched the Junctions howling at the nearby 8 Mile pack. 1047 had been born into that family before switching to the Prospect pack. As the two packs traded howls, a college student named Claire who was with me, the daughter of an Alaskan friend, got so excited she said, "This is the best day of my life!"

I often did talks on the Yellowstone wolves for school groups and park visitors. One spring students from Utah's Pacific Heritage Academy came to Lamar Valley. I spoke to them about the success of Yellowstone's Wolf Reintroduction Project and told them stories of famous wolves. After I answered some of their questions, the teacher told me the students, who looked to be junior high school age, wanted to do something for me.

Their gift turned out to be a traditional Polynesian haka dance. One of the thirteen-year-old boys led a chant and the dance, which I later learned was originally performed to intimidate an enemy army. The chant was very loud and the students, both boys and girls, marched toward me while voicing it. It was easy to see how that could scare off a rival tribe. I ended up speaking to groups from that school many times over the years, and each time I got to see the haka dance. After those experiences, I thought how loud group howling from a wolf pack could serve the same purpose as that chant, in the sense that it could scare a rival pack and cause them to leave without putting up a fight.

969 WAS NOW regularly going in and out of the Natal Den, a sure sign she was about to have her pups. 907 was also pregnant and I guessed she would use the other den at Slough Creek, the one we called the Sage Den. Black Female frequently went into the Sage Den, which made me think she was pregnant as well. Black Female and 907, who were niece and aunt, often squabbled at the den entrance, but not enough to stop them from sharing the den when the time came for them to have their litters. 969 remained uphill at the Natal Den. We think the three mother wolves had their pups around the middle of April.

I saw 1047 come back from a successful hunt one day and give pieces of meat to both 907 and 969. Other pack members did the same or regurgitated meat to the three females. We noticed that 1109 also looked pregnant. She was based on the far side of Specimen Ridge, the area where 970 and 907 had denned in earlier years. It was where 1109 had been born to 970, so she knew the site well. That location minimized her contact with the other females who were aggressive to her. My impression was that she was not willing to put up with their bullying. I respected her for that.

Male 1048 was often seen at 1109's site and was apparently helping her. I was rooting for 1109 and glad that the big male was on her side. He seemed to be her primary support system. That indicated he probably was the sire of her pups. As far as I could tell, he was the only male assisting her on a regular basis. 1048 also visited the dens at Slough Creek, so he was bringing food to the three mother wolves there as well. I got the impression he was especially attuned to what was going on with the females in the pack. I had always admired 1048, and his actions that denning season increased my respect for him. He seemed to have no ambition to take over the alpha male position and appeared to stand outside the pack's male hierarchy.

There were still patches of snow in the den area and the mother wolves would frequently come out of their dens, gulp down mouthfuls of snow to fuel their milk production, then go back into the den to nurse their pups. Alpha male 1047 seemed to be especially taken with 969's pups up at the Natal Den. One day he went into the den tunnel eight times to check on her litter. I recalled he had mated with her at least four times, so it was highly likely all the pups were his.

IN MID-APRIL, TEN 8 Mile wolves suddenly showed up at the Junction pack's den. They spotted 1047 in the distance and ran toward him. He led them north, away from the den. But the rival wolves turned around and headed back toward the den area. Five of them went up to the Natal Den entrance and sniffed around. Two went partway into the den at different times, but both backed out right away, likely because 969 was inside growling and lunging at them. She would be the last line of defense between them and her pups.

Those wolves seemed to lose interest and trotted downhill. But soon most of the 8 Mile pack went back up to the Natal Den. Some went a few feet into the tunnel. They backed out and wagged their tails as they sniffed around. Suddenly they all turned and looked northeast, toward where I had last seen 1047. I guessed he must have howled from that direction. That distraction seemed to change whatever agenda the 8 Mile wolves had, for they all suddenly walked off from the den and left.

I tried to figure out why they left without doing any harm to the mother wolves or pups. What came to mind was that some of the older 8 Mile wolves might have recognized the scent of 1047, a former pack member. That could explain why they were wagging their tails while sniffing around. I later checked my records and found that 1047 had dispersed from the 8 Mile pack three years earlier. We think wolves can recognize howls from fellow pack members. That likely would apply to former pack members as well. I wondered if the combination of 1047's scent and howling was the reason the rival wolves left the den site without trying to mount an attack.

That was the second time I had seen the Junction wolves survive a potential attack on their denning females and pups. The earlier one was in 2016, when the Mollies wolves had shown up at the Junction den area. In both cases, male Junction wolves tried to lure the rival pack away from the den. Despite the rivalries among the females for rank, the Junction wolves cooperated well with each other when hunting, when defending the territory from rival wolves, and especially when raising pups.

As I thought about the times of cooperation and times of aggression in the pack, a famous passage from the book of Ecclesiastes in the Old Testament came to mind. It reads: "For everything there is a season, and a time for every matter under heaven: a time to be born, and a time to die; a time to plant, and a time to pluck up what has been planted; a time to kill and a time to heal... a time to love and a time to hate; a time for war, and a time for peace." For now, it was a time of relative peace at the Slough Creek den. But that was about to change.

Later that month, 1109 traveled to the Slough den area but the other females made it clear she was not welcome. 907 went after her and the two females got into a fight. I was impressed with 1109, for she was holding her own against the bigger 907 and even kept her tail raised, a sign she was not backing down. But then Black Female came out of the Sage Den and ran over to where the other two were fighting. She teamed up with 907 and they chased off 1109.

1109 was later seen crossing the park road to the south, heading back to her own den. After seeing the way the other females were treating 1109, I understood why she denned apart from them.

1109 AND BLACK Female were closely related, but they had totally opposite personalities. I had already classified 1109 as a Maverick and now it looked as though Black Female, based on the increasing aggression she was directing at second-ranking 907, was trying to climb higher in the female hierarchy, a clear indication that she was in the Rebel category.

The tension between Black Female and 907 continued. One day she approached 907 with a raised tail and chased her. 907 ran off with her tail tucked. Later 907 went into a submissive crouch and dropped to the ground under Black Female. There were times when the younger wolf tried to block 907 from going into the den they both were using. They continued to snarl and snap at each other at the den entrance.

Around that time, I noticed 969 limping badly on her right front leg. When she walked, she tried to hold that paw off the ground. We think her injury was a major reason for what happened next. 907 and Black Female, despite their recent squabbling, formed a coalition and overthrew 969. That was a startling event, for we knew that 969 was the mother of Black Female. Then 907 surprised us all by standing up to her niece who had been dominating her, defeated her in a fight, and reclaimed the title of Junction alpha female.

I thought more about the two older sisters and their history. 907 had been the pack's alpha female in previous years. Then, in the spring of 2017, 969 took that position from her. Now, a year later, 907 was taking it back. I guessed that sooner or later Black Female would challenge 907 for the position.

Despite the drama surrounding the three females, we next saw 969 carry her pups to the Sage Den. From then on, all three mothers went in and out of that den as they cared for their pups. The conflict among the females receded as they worked together as a team. That included cooperative nursing, when mothers nurse pups born to other females. It was a startling sight to see that happening after the three females had been so competitive with each other, but that case history shows how rivalries can be put aside for the good of the pack. Referring back to the biblical passage, for the Junctions it was a time for pup rearing. In May we had a high count of eleven pups at Slough Creek.

THOSE OF US who knew the Junction wolves well worried about 1109 and her pups at their isolated den on Specimen Ridge. She looked very thin that denning season, even with 1048 helping her.

The pups at Slough Creek seemed to be doing well. When they were twenty-one days old, I saw them wrestling with each other and playing other games. One pup pulled the tail of a mother wolf. That play was a sign they were healthy and in good spirits.

When the pups were four weeks old, 1109 once again visited the other wolves. She was carrying what might be considered a peace offering: a big elk leg. Despite that, 907 pinned her, but Black Female left her alone. Later, when 1109 started to head back to her den, most of the adult wolves followed her. 969 was the only mother who stayed to watch over the pups. We assumed the adults that went with 1109 met her pups and we hoped that would motivate them to bring them food on a regular basis.

1109 returned to Slough Creek later in May. She went down into the Sage Den to visit the pups. Later they came out and followed 1109 around. At one point, 907, 969 and 1109 were all in the Sage Den with the pups at the same time.

Black Female was nearby and did not harass 1109 that day, nor did any of the other females.

By the end of May, the females were gradually weaning their pups. On May 28, one of the mothers was seen carrying off a dead black pup, and the next day Black Female buried a gray pup. Only three live pups were glimpsed that day and all of them were black. We worried that the problem was distemper. As mentioned earlier, black pups have a better resistance to distemper than gray pups. Now that milk production was coming to an end, the mothers were no longer transmitting antibodies in their milk that could help their pups survive distemper. That meant the pups were at their most vulnerable stage.

When I checked on the Slough Creek den early on the morning of May 30, all the adults were gone. None were there the following day either. We spotted the main Junction pack occasionally in June and July, but no pups were ever seen with them. I later heard that Wolf Project tracking flights spotted three pups (two blacks and a gray) a few miles west of Slough Creek, in an area that was not visible from the road.

ON THE MORNING of August 2, I was looking for wolves at Slough Creek and spotted 1109 coming into the area with those three pups: two blacks and one gray. At the time, we assumed she was the mother of the pups, but we later came to think that they had been born to another Junction mother. Whatever her relatedness to them, it looked like 1109 was devoted to caring for those pups.

The next morning, 1109 and the pups were still in the Slough Creek area and she played with them. The thought came to me that 1109 was acting as their nanny. The pups were very active and rambunctious, exactly how healthy pups should be at that age. In the following days, other adults from the pack joined that small group. We saw that the biggest pup, a gray, was a male. He had a distinctive light stripe down his back. One black pup was female and the other was male. The gray pup was often the leader when the pups were away from the adults. The black female was later collared and became known as wolf 1276. She grew up to be a high-ranking wolf in the pack.

907 joined 1109 and the pups near Slough Creek on August 4 and the two females got along well. That day we saw 1109 playing with the pups for a long time. At one point, the two black pups ganged up on the gray male and chased him. Later the pups moved off to the north and both adults followed them. Soon after that, all eight Junction adults were with 1109 and the pups in Lamar Valley. They ended up at the Chalcedony Creek rendezvous site, a central location used by other packs in Yellowstone going back to the 1990s.

In late August, when the Junction wolves were at the Chalcedony site, a big bull bison moved toward the pups. Both black pups walked away but the gray pup defiantly stood his ground. I thought the pup might have the strength of character to be a future alpha male. Later all three pups followed the huge bull around the meadow.

The Junction adults left to go hunting and later brought meat back to the pups. A black pup got a piece of meat from 1047 but the gray pup stole it. Later Black Male arrived with some meat and tossed it playfully in the air. One of the black pups ran over and grabbed it when it landed.

The Lamar Canyon pack in 2013, after the death of founding alpha female 06 and the dispersal of alpha male 755. Alpha male 925 places his head over the shoulders of alpha female Middle Gray. 926 is the black wolf on the left.
Jeremy SunderRaj

The Lamar Canyon pack in late 2016 after several males from the Beartooth pack joined 926 and her daughter Little Tornado. Those new males were known as 949, Husky Black, and Small Dot. **Jort Vanderveen**

The Lamar Canyon wolves in late 2014 monitoring a cow elk they had injured. She is using a pool of open water in a frozen creek as a defensive position. Due to the site's remote location, we do not know the outcome of the standoff. **NPS/Dan Stahler**

949 by a bison carcass in 2017, a few days before he died. 949 had distemper, which is highly infectious. In his final days, he stayed away from the pack, stoically enduring his fate. **Julie Argyle**

A Lamar Canyon female yearling helping two young pups cross the park road near the family's den. **Jeremy SunderRaj**

Six-year-old Lamar Canyon alpha female 926 in the spring of 2018.
Jort Vanderveen

926's father and founding Lamar Canyon alpha male 755 (left) with a new mate (right) in 2015. Together they founded the Wapiti Lake pack. By that time, 755's once-black coat had turned gray. **NPS/Douglas W. Smith**

The Junction Butte pack in late 2015. The lead gray is 969, who vied with her twin sister 907 for the position of alpha female. 969 is followed by beta male 890, who later left the pack to join the Mollies. **NPS/Erin Stahler**

907, daughter of 911, was born into the Junction Butte pack in 2013. As of early 2023, she has reigned as the pack's alpha female three times and has had more litters than any other female Yellowstone wolf. **NPS/Jeremy SunderRaj**

Junction alpha male 911 in 2015. He died the following year when he heroically fought a rival pack by himself, likely saving his family from a fatal attack. **Jim Futterer**

Longtime Junction alpha male 1047 in 2019. Born into the 8 Mile pack, he dispersed to the Prospect Peak pack and later to the Junction pack. 1047 became the family's alpha male after the death of 911. **Russ Kehler/ Old Wolf Photography**

Black Female (later known as 1382) in 2020. She had a long-term rivalry with her aunt 907 for the Junction pack's alpha female position. **Jort Vanderveen**

Junction female 1276 in April 2022, soon after giving birth. That year, 1276 allied with 907 against alpha female 1382. The pack sided with 907, restoring her to the position she had held before she was ousted by 1382. **Jim Futterer**

Fourth Mother in April 2022, carrying one of her pups across the Lamar River Bridge on her way to a den site where 907 and 1276 had moved their pups. The three Junction litters quickly integrated and played together. **Jim Futterer**

Junction adults and pups at the pack's rendezvous site in Lamar Valley in August 2022. The gray wolf on the far left is nine-year-old alpha female 907. Under her experienced leadership, the pack had fifteen surviving pups at the end of the year. **NPS/Nikki Tatton**

1048 in 2023 followed by a pup. He dispersed from the Prospect Peak pack to the Junctions and later to the Mollies. Although never driven to be an alpha, he sired more known pups than his close relative Junction alpha male 1047. **Julie Argyle**

The adult wolves took the pups to an old bison carcass and the whole family fed on it. When a grizzly arrived, the adults and pups chased it off. The pups followed the big bear for a while, then got bored and returned to the carcass. When that carcass was consumed, the pups occupied themselves by hunting for voles in a nearby meadow. I remembered watching Druid pups doing the very same thing in that same meadow twenty years earlier.

I THOUGHT A lot about 1048 at that time, for he was often with the pups and adult females at the rendezvous site. He was a big, strong wolf with an easygoing personality who got along well with the other wolves in the pack, both male and female. Both 907 and 969 chose to mate with him. I was intrigued by that and wondered if the females sensed he could be counted on to support them during the denning season.

We could see that Black Male was the second-ranking male. 1047 regularly jumped on the backs of males 996 and 1048 and pinned them. But, as before, when he did that to Black Male, the younger wolf did not go down. I also saw that when alpha female 907 did a scent mark, Black Male often marked over the spot. I figured that someday he would challenge 1047 for the alpha position.

In September the Junctions still were based in Lamar Valley but were never seen acting aggressively to the smaller Lamar Canyon pack, which was still splitting its time between the valley and the area around Silver Gate. Early that month, I saw nanny 1109 carry an elk leg to the three pups and give it to them. I also saw Black Female regurgitate a pile of meat to the pups. The three pups continued to play together a lot, a sure sign of good health. Often the play involved stealing food from each other. A week later, two of the pups played a chasing game and both alphas joined in. Those were good days for the pack.

In October the three pups were left alone at the rendezvous site for three days. They did a lot of howling in what seemed to be an attempt to contact the older wolves. A black pup chased a bull elk and three cow elk for some distance. That was a sight: four big elk running from a miniature wolf.

On the third day, the pups traveled on their own west through Lamar Valley. After going about seven miles, they ended up south of Slough Creek. The next morning, I found the pups and all the adults reunited back at the Chalcedony rendezvous site. Later that day, the pups started to chase each other and alphas 907 and 1047 joined in on the games.

The Junctions, a pack that numbered eleven adults and pups, were now ranging from Lamar Valley to the Hellroaring Creek area, an east-west distance of twenty miles. That was almost as large a territory as the Druid wolves had at their peak of power in the early 2000s, when legendary wolves 21 and 42 were the alpha pair.

IN DECEMBER THE Wapiti pack came north from Hayden Valley and I saw them chase a cow elk. White Female, former mate to 755, easily caught up with the cow and grabbed the side of her throat. Another wolf also bit into the throat. Working together, the two wolves pulled the elk down. Then the rest of the pack ran in and helped finish the cow off.

As the wolves fed, I remembered that White Female had been sniffing at a game trail in the snow. The wolf had followed that trail with her nose to the ground. Other elk were in the area, but she stayed on the scent trail. After about ten minutes, she spotted the cow elk just ahead of her on that trail and charged at her. Healthy elk can outrun a wolf pack, but this cow was quickly caught. The white wolf must have detected a particular scent in the trail that told her this elk had something wrong with her, perhaps an infection.

Recently I learned that dogs have been successfully trained to detect Covid-19 infections in people. I was intrigued by that, so I went online and found a site sponsored by the In Situ Foundation. They train dogs to sniff out diseases such as cancer. I learned that cancerous cells give off a scent and that dogs can be trained to identify that odor as well as the scent of other diseases. Their sense of smell is estimated to be ten thousand to one hundred thousand times better than humans'. After training, one of the dogs can smell ten samples in just thirty seconds and indicate which ones are cancerous. That ability comes from their wolf ancestors, wolves like the Wapiti alpha female.

After that successful elk hunt, I heard that Wolf Project biologist Matt Metz had analyzed our records and found that in an average year Yellowstone wolves kill between 5 percent and 10 percent of the park elk. In my years of watching wolves hunt elk, I have seen a pattern where the healthiest elk can easily outrun a pack of wolves. That strongly suggests that wolves remove the least fit elk from the population, leaving the strongest members of the herd to breed and raise the next generation of elk.

As 2018 ended, I reviewed all the things the Junction wolves had gone through during the previous twelve months, and I realized that the lowest-ranking adult, 1109, had served the pack well. Thanks to her being such a devoted caregiver, the Junctions could start the new year with three healthy pups.

9

Coup d'État: Black Female's Story

DURING THE 2019 February mating season, alpha male 1047 tied with Black Female three times and 1048 mated with 907. We did not see 969 or 1109 mate but apparently both did. As the denning season approached, 1109 went back to her den on the south side of Specimen Ridge, near Agate Creek. 907 and Black Female were based at the Slough Creek den and both of them had pups there. We later found out that 969 was denning a few miles east of there but was often at the Slough Creek site.

At that time, 907 was still the alpha female. 969 and her daughter, Black Female, were going back and forth over who was next in the hierarchy. In early May 2019, before the new pups had come out of the two dens, Black Female and 969 formed an alliance and viciously attacked 969's sister, 907, injuring her to the extent that she lost her alpha status. The extreme violence of the attack on 907 caused me to think she would never again be an alpha.

We assumed that 969 would once again take over the top position, but Black Female surprised us when she turned on her mother, just as 907 had turned on Black Female the previous year when the pair overthrew 969. It had served the interests of Black Female to form that alliance with 969. Once the two of them defeated 907, however, her interests changed and she attacked the wolf who had raised her. All that confirmed Black Female was in the Rebel category and gave us more proof that wolves, like people, are capable of betraying allies, even when one is a parent. At the time, Black Female was nearly four years old, while 907 and 969 were six.

After the attack, 907 appeared to be in a lot of pain and looked so bad I was worried she would not survive. Alpha male 1047 often licked her wounds for long periods. That would clean the injured areas and help to heal the sites, for wolves have an antiseptic component in their saliva. Beyond those medicinal aspects of the licking, 1047's attentions to 907 likely calmed her and let her know that she had supporters in the family. It had been very hard for me to see the attack on 907 and witness her suffering, but seeing 1047 helping her reassured me that she was going to get through this.

Wolves' cooperative behavior and their care for wounded companions have roots that go back thousands of years. In the fall of 2021, I made a trip to Los Angeles to do some talks in connection with the release of my book *The Redemption of Wolf 302*. Blaire Van Valkenburgh, a friend who is a professor at UCLA, took me on a tour of the La Brea Tar Pits. Blaire studies the animals that have been exhumed from the tar pits. Those Ice Age species include the gray wolf and the now extinct dire wolf. The staff estimates the animals lived there a minimum of eleven thousand years ago.

Blaire told me that she and her colleagues found remains of many individuals from both wolf species with major life-threatening skeletal injuries that had healed. Those wolves often lived for many more years. The researchers concluded the wolves survived because other pack members fed and cared for them while they were incapacitated.

Later that year, we discovered something that made all the infighting among the Junction Butte females even more interesting. The Wolf Project brought in a helicopter so that some new wolves could be darted and radio-collared. The crew recaptured 907 as the battery in her collar was running out of power. In addition to putting a new collar on the wolf, they checked her for any health issues, such as broken bones and other injuries. We had earlier noticed that her left eye did not look normal. Now the crew discovered that 907 was blind in that eye. She may have been stabbed there by an elk antler during a hunt or bitten in the face during a fight with a rival, probably Black Female. Despite being one-eyed, 907 was the pack's alpha female until recently.

WE FIRST SAW pups at Slough Creek on May 8 and got a preliminary count of three blacks and a gray. They all seemed healthy. Eventually we got a count of seven pups.

969 often traveled to the Slough Creek den site, meaning she was visiting pups born to 907 and Black Female. One day I saw her do something bizarre. 969 went into the Sage Den and came out with a pup that seemed very lethargic, a

possible symptom of distemper. After walking off a short distance, she consumed the sick pup. I had never seen a wolf do anything like that. To this day, I am unsure why she did it. Was it to prevent the sickness from spreading to other pups? Or was it getting revenge on a rival female by killing one of her pups? The previous year, 907 and Black Female had overthrown 969 from the alpha position, and that pup would have been born to one of those mothers.

969 left the site where she ate the pup and headed back to the Sage Den. Male wolf 996 was moving downhill with a pup in his mouth. He dropped the pup and 969 carried it out of sight into a gully. Both that pup and the earlier one were moving a bit, so they were alive. 996 carried a third pup downhill into the gully and it also was alive. After that, 969 carried a fourth pup into the gully. Then she carried what seemed to be a dead pup out of that area and buried it. In the next few minutes, 969 carried three more pups out of sight to the north. We could not tell if they were alive or dead.

In the following days, I watched the Slough den area closely and never saw any live pups. I reluctantly had to conclude that all seven pups were dead and that 969 had likely killed more than the one I saw her eat. We eventually found out that the Junction wolves were now based at 969's den, where she was raising seven pups (six blacks and one gray). None had any signs of sickness and all seven survived the denning season.

907's radio collar recorded GPS positions and the data showed that she was often at 969's den. But she also visited 1109's den several miles to the south. We had sightings of other Junction wolves going back and forth between the two dens. I saw 1048 heading in the direction of 1109's den and remembered that he had helped 1109 at her den the previous year.

907 had been attacked by 969 and Black Female in May. By late July, I saw that all her wounds had healed, and fur had grown over those spots so thickly you could no longer see the injured sites. It was an amazingly fast recovery.

THAT SUMMER I paid special attention to a Junction yearling we were calling Gray Male. I could see that he was already a big wolf. One day he snuck up to a huge, bedded bull bison, twenty times his size, and poked the bull on the hip with his nose. It got the giant animal up. Why did the young wolf do that? Maybe he was just teasing the bull, but more likely the wolf was testing the vigor of the bison.

July 31 was a significant day for those of us who were rooting for 1109. I spotted her on a ridge south of Slough Creek and she had three new black pups with her. After losing her pups last year, it now looked like 2019 was going to be a good year for 1109. We later determined through DNA testing that 1048 had fathered at least one of 1109's pups that year: a black female known as wolf 1229.

A few days later, I spent my eight thousandth day out in the field in Yellowstone. That was about equal to twenty-two years. I saw wolves on 94 percent of those days. Soon after that date, I typed my twelve thousandth page of field notes on my computer. Since 1995, the first year of the wolf reintroduction, I had gone back to my cabin every night after watching wolves out in the park and typed up what I had seen.

IN AUGUST THE seven pups at Slough Creek were moved to the Chalcedony Creek rendezvous site on the south side of Lamar Valley. Later that month, two big bison died a few miles to the west. The adults brought the pups to the carcasses and the pack fed for many days. The wolves had a lot of competition. One day I saw five adult male grizzlies along with a mother grizzly and two cubs, a total of eight bears. The pups quickly learned how to feed at spots away from the bears.

I saw a Junction yearling dig a hole, drop a piece of bison meat into it, then push dirt over the spot. All that is the normal procedure when a wolf wants to hide food for later consumption. But it would be easy for other pack members to notice the fresh dirt. A wolf could guess that food had been buried there and steal it. This yearling did something new. It plucked some grass and laid it down over the site. The behavior looked like an attempt to hide the fresh dirt from the other wolves.

I contrasted that incident with a later one when a young 8 Mile pup dug a hole, dropped a piece of meat in it, then covered it up with just dirt. Three other pups watched the covering-up job from a few feet away. When the naive first pup walked off, one of other pups went over, dug out the meat, and carried it off.

As I thought more about the Junction yearling's attempt to hide the meat from other pack members, I remembered a time when I was watching the Wapiti Lake pups in Hayden Valley. The mother wolf, her yearling daughter, and the three Mollies males had left the four-month-old pups at a rendezvous site and taken off on a hunt. I later saw them coming back toward the pups from the south. The adults howled and the pups howled back. Three of the four pups looked around, then stared south, the correct location of the adults. But the fourth pup seemed confused about where the howling was coming from. Rather than continue to scan around in various directions, that fourth pup looked at its three siblings. It must have seen that they were all staring south, for it immediately looked that way. The adults soon trotted in from that direction and rejoined the four pups.

In recent years, there has been a lot of interest in trying to discern if animals are intelligent enough to have what is known as a Theory of Other Minds. That is an understanding that other individuals, both of your own species and of other species, have minds that work somewhat like yours. When the Wapiti wolf pup could not figure out where the pack's adults were when they howled, instead of wasting more time trying to spot them, it turned to look at its companions. After seeing them staring south, the pup must have understood that they were seeing the adults that way, for it immediately looked in that direction. It was an intelligent way to solve a problem.

Pups, like adult wolves, need to figure out smart ways of doing things if they are to survive. The story of the Junction yearling who hid a piece of meat in a hole, covered it up, then laid some plucked grass over the fresh dirt, indicated that the young wolf seemed to understand what might fool another wolf and keep its meat safe for later use. The fresh dirt could tip off a sibling that something edible was hidden there, but the plucked grass covered up the dirt and likely kept the meat safe.

While considering the intelligence and problem-solving abilities of wolves, I recalled an incident where two young Junction wolves had to quickly come up with a solution to a challenging situation. The siblings had a bison calf they were trying to feed on while surrounded by a large herd of bulls and cows. Every time the wolves approached the calf carcass, a group of adult bison would charge in and drive them off. Finally, the wolves figured out a solution. They grabbed opposite ends of the calf, then ran off with it. They had some difficulty jointly carrying the calf, but their plan worked. The two wolves got far enough away from the herd, put the calf down, and were able to feed without any more interruptions. I was impressed by that spontaneous teamwork. It showed that when faced with a problem, wolves are capable of coming up with innovative solutions on the fly.

Then there was the time a Mollies female wolf had a bull elk carcass south of the park road. A big crowd of people had gathered, hoping to photograph the wolf when she returned. I was down the road to the east. Looking toward the crowd, I saw the wolf up high on a ridge to the north. She would have been in plain sight to anyone who turned around, but no one did. After looking at the people between her and the carcass, the wolf went east and disappeared behind a ridge. A few minutes later, she reappeared, rushed downhill, and crossed the road without anyone in the crowd noticing her. They were all concentrating on the carcass. If she had tried to cross closer to the people, some of them would likely have rushed toward her to take pictures and that would have turned her back. It appeared that she had figured out that if people spotted her, they might block her access to the carcass. The wolf came up with a plan to fool the people and it worked. All these incidents implied that wolves are thinking, rational beings.

THE WEEK THE Junction Buttes found the two bison carcasses, 1109 brought her litter to Lamar Valley and they met up with a few of the Junction adults. I noticed that one of her three black pups was smaller than the others and seemed shy around these new adults. I then saw some of those adults share food with 1109's pups. That probably reassured the small pup that these new wolves were friendly.

Soon after that, 1109 and her three pups joined the entire Junction pack, including the other seven pups. Now we had all ten pups together: nine blacks and one gray. The new arrivals integrated with the other pups very quickly and all ten spent a lot of time playing together. There were eleven adults in the family, so the total membership was twenty-one.

The pups were now fearless when they encountered bears. One day in September, a mother grizzly with two cubs came near some of them. The pups charged and the bears ran off. Later three pups helped a yearling wolf get a pronghorn antelope, the fastest animal in Yellowstone, capable of running up to sixty-five miles per hour. The top speed of a wolf is around thirty-five miles per hour, but it looked like the pronghorn was confused when it saw four wolves, somewhat spread out, charging at it. It cut away from the nearest wolf but was caught by other wolves who were running parallel to the first wolf.

That fall I saw three pups run up to 1109 and beg for a feeding. She lowered her head and regurgitated a big pile of meat. The pups scarfed down every morsel. One of the pups was gray and 1109 had no gray pups that year. The pup must have been born to one of the other mother wolves in the pack. That seems to be the pattern among mother wolves: they nurse or feed whichever pups run to them, regardless of which ones are theirs.

My first wolf book, *The Rise of Wolf 8*, came out that October, and that month I went to a recording studio in Bozeman to do interviews for an NPR show. On the walls of the studio were photos of celebrities who had done interviews there. I snuck out before they could take my picture.

There was a terrible tragedy one night in November. Two of the black Junction pups were hit and killed by a car. That changed the membership count to nineteen, which was still twice the average size of a Yellowstone wolf pack. There were eight pups now.

On January 12, 2020, we had the twenty-fifth anniversary of the arrival of the first batch of reintroduced wolves from Canada. Those eight wolves were the original members of the Rose Creek and Crystal Creek packs. The Crystal wolves later were renamed the Mollies pack. The females that founded the Junction pack were born into that family. The founding males were from the Blacktail pack and they were partially descended from Rose Creek wolves. That means that descendants of the Crystal Creek/Mollies and the Rose Creek wolves were living within a few miles of where their ancestors had been released in Yellowstone twenty-five years earlier.

That month 969's health declined. She died at Slough Creek, not far from the den area, and weighed only seventy-three pounds when examined. In contrast, her sister 907, who had been recollared a few weeks earlier, weighed 121 pounds. 969 had a puncture wound from a bite that went all the way into her abdominal cavity. We suspected the injuries had been inflicted by another Junction female. If that was the case, it was probably related to the incident the previous spring at the Slough den when 969 killed and consumed a young pup and likely killed additional pups.

The situation reminded me how, back in 2000, an alliance of Druid females attacked and mortally wounded wolf 40, the pack's alpha female, who had previously killed two litters of pups born to her sister. In both cases, a mother wolf exhibited aberrant aggressive behavior toward pups born to other mothers and it appeared that the pack later killed the offending wolves.

907 WAS NO longer the pack's alpha female but she often led the pack when they traveled. The Junctions now numbered seventeen with seven pups when 1109 and her litter were in the group. In the February 2020 breeding season, 1047 mated with 907 and later I saw her in a mating tie with Black Male. We know that wolf pups in one litter can have two different sires, so it was likely that both males would father pups born to 907 in the coming spring.

The snow was deep in our section of the park at the time. I saw sixteen Junction wolves traveling single file through a trench in the snow that had probably been made by a bison herd. I could just barely see the backs of the taller wolves. Using that trail was a smart thing to do, for it saved the wolves a lot of energy.

Doug Smith, head of Yellowstone's Wolf Project, once did a talk I attended and discussed something called surplus killing. That refers to stories of wolves killing more than one prey animal at a site and supposedly leaving without eating much of the meat. These stories are repeatedly told by anti-wolf people to justify the killing of wolves.

I am writing this chapter in the spring of 2023. I have now been in Yellowstone for twenty-nine years and have been in the field for well over nine thousand days. In my time here, I have seen Yellowstone wolves make 210 kills over the course of 205 successful hunts, where wolves killed medium-to-large-sized prey animals such as elk, bison, mule deer, bighorn sheep, and pronghorn. The wolves got just one animal on two hundred of those hunts. On five hunts, the wolves made a double kill. In three incidents it was two elk calves, and on two other occasions they got two cow elk. Three of those double kills (two sets of cows and one pair of calves) were in late winter when elk are in their poorest condition. The other two pairs of calves were taken in June when weaker calves have a hard time outrunning wolves.

In the first late-winter case, fourteen wolves fed all day (twelve hours) on two cow elk kills. Deep snow that morning enabled the wolves to easily catch up with both cows and pull them down. During a second late-winter hunt, seven wolves approached two cow elk in midmorning. One cow, which must have been in very poor shape, did not try to run off and was quickly killed. The other cow did run but the wolves easily caught her. The wolves were still at the two carcasses when it got dark, over ten hours later.

The third late-winter hunt involved two elk calves that were about ten months old. Both were undersized for their age, an indication they were not in good health. The first calf ran into a field of deep snow, got bogged down, and was taken down there. The other calf ran into that same area of deep snow and foundered at the exact spot where the other calf had been caught. The wolves then fed on the two calves.

To sum up, in 97.56 percent of the 205 successful hunts I have witnessed, a single prey animal was taken by wolves. In just 2.44 percent, the wolves took two animals. In all cases, the wolves ate what they successfully hunted.

I later had a good chance to see how fully wolves consume a carcass. The Junction wolves were feeding on the well-picked-over remains of a bison carcass. One young female carried off a large section of the pelt. She put it down, plucked off the fur, then ate the hide. A pup joined her and also ate some of the hide. Another young female gnawed on a horn and a different wolf carried off the skull. It was likely also going to gnaw on that. A few months later, I saw Junction wolves eat parts of a hide at another carcass. Previously I had watched 926 eat a hide and seen wolves eat the lining of the stomach. All that confirmed to me that if humans stay away from a kill site, wolves will consume it down to those unappetizing parts. In contrast I once saw a report that found humans throw away a third of our still-edible food. Based on that fact, we are not well suited to accusing other species of being wasteful.

Another accusation that often comes up is that wolves kill too many elk calves. A study in the early years of the Wolf Project kept track of elk calf predation. As described in the 2020 book *Yellowstone Wolves: Science and Discovery in the World's First National Park*, this study found that grizzlies and black bears kill about four times more elk calves than wolves do.

A related issue is the number of elk that wolves take each year compared with the total elk population. I sometimes hear accusations that the wolves "are killing all the elk in the region." According to the Montana Department of Fish, Wildlife and Parks, in 1995, the year of Yellowstone's wolf reintroduction, the estimated number of elk in Montana was 109,500, and in 2018 it was 138,470 despite 27,793 elk being shot by hunters that year.

The Wyoming Game and Fish Department estimated that the number of elk in Wyoming in 1995 was 103,448, and in 2018 it was 110,300. In 1995, hunters in that state shot 17,695 elk, and in 2018, twenty-three years after the wolves were brought back, they shot 25,091.

In 2021 the elk counts in our three local states (Idaho, Montana, and Wyoming) were 40,000 to 60,000 higher than the 1995 levels, according to former US Fish and Wildlife Service biologist Carter Niemeyer, who reviewed reports from those states. Although people who were against the reintroduction predicted that wolves would "decimate the elk population," the elk population increased significantly after wolves were reintroduced.

The Montana Department of Fish, Wildlife and Parks announced in the summer of 2022 that the state's elk population has tripled over the last four decades. The agency's goal for statewide sustainable elk numbers is 92,000, but the current estimate of elk in Montana is 175,000. That means the state elk biologists say there is an overpopulation of 83,000 elk in Montana.

Another finding from Wolf Project research on elk is that the average age of cows taken by wolf packs is fourteen years, while cows shot by hunters have an average age of 6.5 years. That would be in the middle of a cow's prime reproductive years, while fourteen-year-old cows are near the end of that period.

My friend Norm Bishop, a retired longtime Yellowstone biologist and information specialist who periodically sends me statistics on elk numbers from the fish and game agencies in Montana and Wyoming, summed up all that by saying, "Human hunting has far more effect on elk population dynamics than does wolf predation. And wolves take the sick and weak, removing them from the population, strengthening the remaining healthy ones."

IN EARLY 2020, we were looking forward to having another great spring of watching the Junction wolves raise their pups at the Slough Creek den. Black Female, the alpha, and 907 and 1276 all looked pregnant. Their due dates were late April. The pack had eleven adults and eight pups from the previous year. But then a worldwide crisis ended up impacting Yellowstone. Yellowstone closed on March 24 because of the coronavirus epidemic. That meant I could not go into the park, other than to drive to the town of Gardiner for food. We were not supposed to stop for wildlife viewing on the way.

Jeremy SunderRaj, who was still working for the Wolf Project, was allowed to monitor the wolves during that time. He told me that 1109 had dug out a new den near Crystal Creek a mile and a half south of the Slough Creek den, and that her ally 1048 had been seen there with her. Three other Junction females had pups at Slough Creek: 907, Black Female, and 1276. In early May, Jeremy saw six pups come

out of the Natal Den, and 907 nursed them. Later a yearling carried a pup downhill from that den. 907 carried it back
uphill.

WHEN I WAS stuck at my cabin during the park closure,
I worked on my book about wolf 302, organized all the
wolf-related material on my computer and in my cabin, and
walked around town. I got to see a grizzly on one walk and a
cow moose and her yearling during another walk. A huge bull
bison walked by my cabin window one morning.

I started living year-round in Silver Gate in 1999. It is a
small Montana town surrounded by the Absaroka mountain
range. Our elevation is 7,390 feet and nearby Amphitheater
Peak rises to 11,042 feet. The entrance to Yellowstone is a
mile away and a National Forest Wilderness area is a short
walk from the center of town. Soda Butte Creek runs west
through Silver Gate. It merges into the Lamar River, then
flows into the Yellowstone River.

In the early years, I rented a place that had once been a
one-room schoolhouse. I bought a small cabin in 2003 and
am still living in it. Often there are only about seven of us
here in the offseason. We get a lot of snow in the winter and
spring months. It was hard to find accurate information on
just how much, but one source said it averages 194 inches.

The population greatly expands in the summer tourist
season. We have a few businesses and stores: the Log Cabin
Cafe (they serve great trout dinners), Stop the Car Trading
Post, the Grizzly Lodge, the Silver Gate General Store &
Silver Gate Lodging. That last business also runs the Range
Rider Lodge. The lodge is a huge log structure that was built

in the late 1930s and is by far the most impressive building in our small town. It has a bar, rooms to rent, and a big room where some of us do talks for the community and tourists. In earlier years, dances were held in that room.

Like all small towns, we have our stories. Ernest Hemingway fished and hunted in the area and drank at the bar in the Range Rider when he vacationed here in the 1930s. The last short story published in his lifetime was "A Man of the World." It came out in 1957 and was set in a fictional version of Cooke City, the slightly larger town three miles east of Silver Gate. His earlier novel *For Whom the Bell Tolls* was partly written during his time here. He would mail new chapters of the book to his editor in New York City from the local post office. A big sign outside the Range Rider Lodge says, "Drink where Hemingway drank."

Henry Finkbeiner, who owns Silver Gate General Store & Silver Gate Lodging and the Range Rider, is a tremendous asset to the town and a great help to our residents and park visitors. Recently I was talking to a man out in the park and he told me that years earlier all his credit cards were compromised and rendered useless. He was running out of cash and did not know what do to do. Stopping in Silver Gate, he happened to tell his story to Henry. To the man's astonishment, Henry put a pile of money in his hand and told him not to worry about paying him back.

One of the town's residents has the job of keeping me humble. Since my wolf books came out, I often have people asking me to sign their books or be in a photo with them. When I come back to Silver Gate and pull up to my cabin, my neighbor's dog Keeper always runs out into our shared driveway with great excitement, expecting to see Laurie, her owner. But when Keeper sees that it is just me, someone who is useless to her, an expression of great disappointment flashes across her face and she turns around and goes home. That keeps me in my place since it happens pretty much every day.

MAY 13, 2020, was the twenty-fifth anniversary of my seeing my first Yellowstone wolves. It was the Crystal Creek pack, ancestors of the Junction wolves. A few days after that anniversary, I saw 1109 and two other wolves near her den site when I was driving to Gardiner for food. That day, Jeremy told me that all the pups born at the Slough Natal Den to Black Female and 907 were now based at the Sage Den with 1276's pups. He got a count of eighteen pups there. That den must have been huge if it could fit that many pups along with the three mother wolves.

Jeremy saw 907 and Black Female nursing pups. Two-year-old 1276 was lactating as well. With the eighteen pups, the Junction pack now numbered thirty-five. That did not include any pups born to 1109, for we did not yet know if she had any. Only the Druid Peak pack, back in the era of wolves 21 and 42, was bigger. At one point, the Druids numbered thirty-eight. The average Yellowstone wolf pack has ten members.

Jeremy confirmed that 1109 was also producing milk. Later he spotted her traveling with Black Male from the main Junction pack. He scent marked a bush and she marked over it. It was the first time I had heard of 1109 scent marking with a male. That made me wonder if Black Male was the father of her pups.

Back at Slough Creek, Jeremy saw Black Female and Black Male carry pups who had strayed back into the Sage Den. Some of the yearlings also learned how to pick up pups and bring them back to the den when they wandered off. Sometimes a pup would squirm out of an older wolf's jaws and fall to the ground. The wolf would pick it up right away, hold it tighter, and carry it off.

ON JUNE 1, the park reopened after a sixty-eight-day closure. I got up at 3:15 a.m. and drove directly to Slough Creek. I saw pups come out of the Sage Den and got a count of fourteen. I later saw all eighteen. As I watched them, I thought this must be the most exciting time of the year for a wolf pack. Yearling pack members seem especially fascinated with new pups and continually monitor and interact with them. That attention to the pups lets the mother wolves have a break.

By early June, we often saw the little pups wrestling with each other, a sign of good health and physical development. We no longer saw nursing as of June 15. Pups would try to nurse but the mothers would move off and sometimes snap at them.

That month a big cow bison died in Lamar Valley. When I stopped to look, I saw a yearling bison bedded down by the dead cow. The cow was likely the yearling's mother. A black Junction wolf approached and bit the tail of the young bison. It was probably trying to harass it into leaving so the wolf could start feeding on the carcass. Then a big herd of bison arrived, and the wolf had to get out of the way.

A crowd of bison encircled the dead cow and many of them licked her nose, head, and body. More bison came in and did the same thing. Soon there were about a hundred bison gathered at what looked like a funeral. Then the most astonishing thing happened. A bison hooked its horn around a horn on the dead cow and seemed to be trying to help her get up. More bison came in and licked the dead animal as the yearling looked on.

I checked early the next morning and saw that two grizzlies were feeding on the carcass. I scanned around and saw 1047 going west, toward the Slough den. He probably had been feeding there before the arrival of the bears. I spotted 1109, Black Male, and another Junction wolf south of the carcass. They likely had also been at the site. The grizzlies later left the big carcass. Soon after that, 907, 1048, and two other Junction wolves arrived and fed on the bison.

One morning at Slough Creek a man came up to me and said he had just read *The Rise of Wolf 8*. He explained that he had grown up in Idaho and had been a hunter who hated wolves but that reading about wolf 8 had made him think differently and he was now here in the park to watch them. I thanked him and told him that is why I write about Yellowstone wolves.

IN MID-JUNE, 1109 appeared to have moved her pups up the Lamar River. Black Male was still visiting her and they continued to do double scent marks. One day male 1048, a past ally of hers, approached the pair and got in a standoff with Black Male. 1109 tried to lower the tension by doing play bows to the males but they got into a fight anyway. Black Male beat 1048 and chased him off. What a development that was to witness. 1109, who had been picked on by higher-ranking females for years, now had two of the high-ranking males in the pack fighting over her.

Other Junction wolves visited 1109 and Black Male, including females 1276 and 1229, 1109's daughter. We all hoped 1109's relationship with Black Male would result in her rejoining the pack at Slough Creek.

By late August, the Junction adults and their eighteen pups were based at the Chalcedony rendezvous site in Lamar Valley. On September 1, I saw six different adult wolves return to the rendezvous site after a hunt and feed the pups either by giving them meat or by regurgitating it to them. The pack was doing a heroic job in feeding all those hungry mouths.

Soon after that, 1047 injured a front leg during a bison hunt. He limped on that leg as well as on the rear leg that had been damaged back in 2017. He carried on all his duties as a father and alpha male, but Black Male apparently sensed an opportunity and chose that time to challenge him. Even though 1047 was limping on two legs, he defeated the younger wolf and Black Male backed off.

After that fight, we noticed that 1047 had a torn left ear. As 1047 deteriorated in the following days, Black Male ascended to the alpha position. I never saw another fight between the two and concluded that the younger male took over with only a minor degree of dominance toward the older wolf. That put Black Male in the Bider category. I thought of the phrase used in Mafia movies when a younger man takes over the operation: *"It's nothing personal. It's just business."*

1047 had been the Junction alpha male from early 2016 through the summer of 2020. That reign of four and a half years was a long time to be a successful alpha, since the average age of a Yellowstone wolf is only a little over three years. He was now six and a half years old. Black Male was five and strong and vigorous. It was his time now. Now that he and Black Female had taken over both alpha positions, the transition from the older generation to the younger one in the Junction pack was complete.

1047's ear got worse in the following days and by mid-September it was hanging down. He shook his head repeatedly, indicating something was bothering him. I noticed that he tended to tilt his head to the left. His pain did not interfere with his duties in his semiretirement, for I saw 1047 repeatedly feeding the pups. Thinking back about that time, I now realize that Black Male could have driven 1047 off, but he allowed the former alpha to remain in the family. The two males were related and perhaps that was a factor in Black Male's treatment of the old wolf.

1109 WAS BASED a few miles southeast of the main group of wolves by then and seemed to be avoiding meeting up with her relatives. 1109's radio signals indicated she sometimes went east and crossed over the park border. By then months had passed since 1109 had left her den south of Slough Creek. A Wolf Project crew checked the site and found small tracks in the den, proof that she did have pups there. The den tunnel was thirteen feet long and probably originally was a coyote den that she remodeled. We did not know if 1109 had moved her pups to a different location or even if they were still alive. Perhaps her travels to the east and her independent personality were the reason we stopped seeing Black Male visiting her.

In October 1109's daughter 1229 often acted as babysitter for the Junction pups at the Chalcedony rendezvous site. She did a good job supervising them. One day she led the pups away from the rendezvous site on an excursion. 1229 looked like a schoolteacher walking a mob of children to a museum.

That month 1109 visited the other thirty-four wolves in the Junction family and none of them harassed her. After that, she was spotted traveling with a black pup. The following month, she was seen with a black adult wolf east of the park. In December former alpha male 1047 and two black wolves were observed with her. Those were positive events in her life, but by that time we reluctantly concluded that none of her pups from the spring had survived.

Soon after that, we spotted 1109 with five Junction wolves. The group, which was occasionally joined by 1047, traveled far to the west, miles past Slough Creek. Gray Male, the pup from 2018 who 1109 helped raise, scent marked a site, and 1109 and two other wolves marked his spot. We all hoped that meant this was going to be a new pack with 1109 as the alpha female. Despite her past pattern of being independent, she seemed to slip easily into a leadership position and all the other wolves, male and female, were accepting of her in that role.

ON DECEMBER 23, I set a personal record. I saw three wolf packs in the Tower Junction area, about six miles west of Slough Creek. There were twenty-nine wolves in the Junction Butte group, twenty in the Wapiti pack, and another six in a new group known as the Carnelian pack. That added up to fifty-five wolves. A few days later, I also saw 1109 with four former Junction wolves, including Gray Male.

The Wapiti wolves were still in our northern section of the park on December 31. White Female, the onetime mate of 755, was in the group along with former Mollies wolf 1014, who was the pack's alpha male. He played with a pup in a rough manner. The mother ran over and slammed her body into him, like a hockey player would to an opponent. She did that two more times and the male, who was much bigger than she was, did not fight back or even try to nip her. He just took it. That year, 2020, both White Female and another female whose coat was getting very light had pups. If the younger female's coat turned white, she would be the fourth white female in the Wapiti pack.

I always liked seeing this pack when it came north from Hayden Valley, because it reminded me of the tremendous resilience of its cofounder, wolf 755. Every year back then, I hosted a group from Texas Children's Hospital in Dallas when they came to Yellowstone. The young adult counselors were all cancer survivors who were treated at the hospital and the kids were recovering cancer patients. Back in 2017, I took them to Hayden Valley and showed them Wapiti wolf 1091 and her five pups. Then I told them stories of wolves like 926 and 755 (the grandfather of those pups) who, like the kids from hospital, had overcome tremendous adversity in their lives.

AT THE END of 2020, we documented that all eighteen of the Junction pups from the Slough den had survived. That was a spectacular cooperative accomplishment for the pack. Black Female was the alpha female that year, so she deserves a lot of credit for that success. But all three mothers who were based at Slough worked as a team and they had the support of the rest of the Junction pack.

Around that time, I talked with Quinn Harrison, the biologist who studied dispersal among Yellowstone wolves. He told me that cooperative breeding and raising of young is rare in nature. Among mammals it is limited to wolves, spotted hyenas, marmosets, mongooses, meerkats, naked mole rats, and primates.

1109 was still with her subgroup of Junction wolves in early January 2021 and continued to be the highest-ranking female. Gray Male was the dominant male. A few days later, the main Junction group found those wolves and Black Female chased 1109. After that incident, 1109 went back to being mostly a lone wolf, another confirmation of her Maverick nature. To recycle a phrase from the 1960s, being in charge of a pack just "wasn't her thing."

WE SAW JUNCTION Black Male mate with 907 in February 2021. A few days later, he got interested in Black Female. 907 tried to block him from following her but he just went around 907 and continued to pursue the other female. The two alphas got together and mated. After that, 907 went to former alpha male 1047. She averted her tail to him, a sign she wanted to mate with the old wolf.

Younger female 1276 did the same thing with 1047 on a later day. After that, a third female showed interest in him. 1047 was no longer an alpha but the females did not seem to care about his status. He had lived a long life for a wild wolf and that indicated superior health, strength, good genes, and intelligence. In addition, he had proven to be a good provider to mother wolves and their pups in past years. All those would be desirable traits for a female who is looking for a breeding partner.

There was a big development soon after that. Gray Male and another Junction male were spotted with female 1154, a wolf from the 8 Mile pack. Soon more young males from Junction joined them. That turned out to be the start of a new group that became known as the Rescue Creek pack. The alpha pair was 1154 and Gray Male. The group's membership settled out at twelve. Ten were Junction males and two were 8 Mile females: 1154 and a younger female.

1047 gradually deteriorated that winter. He still shook his head a lot, indicating his health was not good. I will always consider him one of our greatest alpha males. The Junction pack prospered during the many years he was in that position, and when his health declined, he seemed to gracefully accept the ascension of Black Male to the alpha role. He later left the pack and went west to the Blacktail Plateau area, where he had lived before joining the Junction pack in early 2017. He died in mid-March. By that time, Junction male 996 had disappeared and likely had also died. That meant there were now just two older males in the pack: 1048 and Black Male.

The following month, Mollies wolves came back up to Lamar Valley. They bedded up high on the south side of the valley. 890 was in the group. He was now a very old wolf, nearly ten. He had successfully made the long trip to the north from Pelican Valley through deep snow, a tough trip even for much younger wolves. That was the last time I saw 890, another wolf I greatly admired.

IN THE SPRING of 2021, three Junction females—907, 1276, and Black Female—once again looked pregnant. Eight-year-old 907 was in good shape for her advanced age. She was especially big for an old female and her size indicated she was still strong and vigorous.

I was always on the lookout for 1109. The last time I saw her was in Lamar Valley with three other wolves, two blacks and a gray, that I did not recognize. They may have been from a pack east of the park. Mark Packila, who does the tracking flights for the Wolf Project, told us that he spotted 1109 east of Yellowstone in late August 2021. She passed away later that year but lives on through our memories of her Maverick life choices and her time as a nanny.

Black Female had now held the alpha female position for two years. She had started out as a Rebel, overthrowing her aunt 907 and then turning on her mother, 969, who was her accomplice in the coup. The pack seemed to be doing well under her leadership.

10

Team Player: 1048's Story

IN APRIL 2021, I spotted eight Junction wolves in Lamar Valley including big male 1048, former alpha female 907, young female 1276, and two pups from the 2020 litters. Both females were pregnant and close to their due date. The Junctions encountered four Mollies wolves: four-year-old gray male 1237 and three younger wolves.

Wolves from the two packs got into fights. I noticed that 1048 tended to stay close to 907 during the hostile encounter. The two pregnant females, 907 and 1276, soon ran off and the two pups followed them, the right thing for them to do. 1048 went with them as well, looking like their bodyguard. The remaining Junction wolves followed his example. The four Mollies wolves did not pursue them, probably because they knew they were outnumbered.

I recalled how often 1048 attended the mother wolves who were denning many miles south of the main Junction

den at Slough Creek. He seemed to be their most depend-
able provider of food, both for them and for their pups.
Without 1048 many of those pups likely would not have
survived. I mentioned earlier that I felt he was especially
attuned to what was going on with the females in his pack.
This incident was more proof of that. The thought came to
me that 1048 was a team player.

SOON AFTER THE incident with the Mollies wolves, 907 and
Black Female prepared to have their pups at Slough Creek.
We saw them going in and out of both the Natal Den and
the Sage Den. A little later, I spotted wolves looking into the
Sage Den and wagging their tails. I assumed that meant they
were seeing or hearing newborn pups in the tunnel. A similar
thing happened at the Natal Den. It looked like Black Female
was using the Sage Den while 907 was uphill at the Natal Den.
1276 had chosen to den in a remote location to the west.

We do not know what happened, but soon we concluded
that Black Female and 1276 both lost their pups. It might
have been due to distemper, but we never knew. That meant
that if the pack was going to have any surviving pups that
year, they would be 907's. It was too early to see pups at her
den entrance but we watched the site every day hoping to
spot live pups. 907 was drinking a lot of water, a sign she was
nursing. We saw yearling wolves frequently going into her
den. Black Male went up to the den entrance and stuck his
head inside. He bedded down and continued to look into the
den. All those activities indicated there were live pups inside.

I first saw 907's pups come out of the Natal Den on May 7.
There were five. A yearling came over and licked them. Then
the pups nursed on their mother. A tiny gray pup tumbled

downhill. 907 ran after it and carried the pup back up to the den. After that, she touched noses with one of her pups. It looked like an affectionate greeting, like a human mother playfully touching noses with her infant. A yearling later carried an elk antler into the den, apparently to give to the pups as a toy.

We saw Black Female frequently go in and out of that den, indicating she was helping 907 care for the pups. 1276 had returned to the Slough Creek den area by that time. Her presence seemed to lessen the squabbling among the two other females. When the pups began to spend a lot of time out of the den, we saw Black Female and 1276 both nurse 907's pups. Black Female had overthrown 907 and taken over the alpha position in 2019, but now that she had lost all her pups, she was devoted to helping 907 with her litter. All three females were now cooperating to make sure the pack's pups survived.

We ended up with a count of five gray pups and three black pups in 907's litter. By the time the pups were a month old, they were walking around well and wrestling with each other. They continued to nurse on all three mothers through the end of May. They were also eating meat that adults carried to the den or regurgitated to the pups. One day two pups had a tug-of-war over a strip of meat. The pup that won ran off to eat in private.

I saw one pup bite the tail of Black Female and pull on it. She did not seem to mind. Another pup crawled up on the back of a yearling and yanked on its ear. When it let go, the yearling wiggled that ear. It looked as though it was tempting the pup to do it again.

JEREMY SUNDERRAJ TOLD me that only older females like 907 seemed to have surviving pups that spring in Yellowstone. A possible explanation was that the older females had been exposed to distemper and survived. That enabled them to pass on distemper antibodies to their pups through their milk. Younger mothers were probably not exposed to distemper and did not have those antibodies.

That month I learned that my second Yellowstone book, *The Reign of Wolf 21*, had won the best narrative nonfiction book award in the Reading the West contest put on by the Mountains and Plains Independent Booksellers Association. It meant a lot because the contest was based on votes from readers.

I saw the Junction pups howl for the first time on June 5. They were still nursing on the three females, but by June 12 the mothers' milk production seemed to be over. I saw Black Female pin former alpha 907 only twice that spring. Perhaps all the energy required to raise the eight pups meant there was little to spare for belligerent behavior.

That month a Park Service naturalist named Julia, who was working in the Canyon area, came up to see the Junction pups. When I showed her the den, Julia told me that she had first met me when she was nine years old. She had fulfilled her dream of one day working in the park.

The pups often followed adults to the nearby Slough Creek but were hesitant about going into the water, even when they saw older packmates swimming across. By late July, the pups were wading out short distances in shallow water. Soon after that, they graduated to swimming. That naturally takes place when they reach deeper water and start to float. The pups seem to instinctively dog-paddle at that point and that gets them swimming. I also saw that the three-month-old pups were already proficient at following scent trails of the adult wolves.

Six of the eight pups were moved to the Chalcedony rendez-vous site in Lamar Valley on August 1. To get there, the pups had to swim the Lamar River, which was much wider than Slough Creek. Mother wolf 907 was with them, so she had probably led the pups there. The distance from the den to the rendezvous site was about eight miles. The two other pups must have balked at crossing the river, for we found them back at Slough Creek. They were south of the den area and no adults were with them. One pup chewed on a bison skull that had been there for well over a decade. Later a pup howled. I felt it was trying to contact the adults.

Two days after that, I saw both pups trotting south, fol-lowing what might have been the scent trail of the adults. They suddenly seemed scared and ran back toward the den. I looked south and saw two men fishing in the creek. That must have been what frightened them. I regarded that as a good thing: to have an instinct to run from humans.

One day there was only one female yearling with the six pups at the Chalcedony rendezvous site. A huge bull bison approached the pups and the young female. The bull, who was more than twenty times her size, charged at her. She deftly darted out of the way, then seized the initiative and countercharged him three times. This bull soon lost interest in the wolves and wandered off. After that, the pups occu-pied themselves by hunting for voles, a favorite activity that served as training for later hunts of bigger prey.

The other two pups remained at Slough Creek. They often howled. On August 4, they went east from the den and reached the creek. One pup swam to the far bank but then swam back, probably because it had seen people in that area. On a later day, I saw the two Slough pups hunting insects and voles. After that, both of them chewed on the same old bone. I watched them closely and got the impression that they were not stressed.

A few days later, I spotted the two pups with their relative Gray Male and the group that would come to be known as the Rescue Creek pack. The following day, they were south of the Lamar River and the park road with several Junction adults, including 1276. They knew her well, for she had nursed them after she lost her own pups. She must have gotten them to swim the river and cross the road. The group traveled east, toward the rest of the pack and the other six pups. On August 12, a group of adults, including Black Female who had also nursed the pups, brought the two pups into Lamar Valley and they had a reunion with the rest of the litter.

Wolf 1229, a daughter of 1109, continued to fit in well with the main group of wolves and was doing a good job of helping to care for the pups. I saw her bringing them food as well as toys such as an elk antler. Another time 1229 stole a cache of meat a grizzly had buried. She carried it directly to the pups and gave it to them. I knew that 1048 was 1229's father. Over the years, I had seen that he had good social skills and seemed to get along well with other wolves. It looked like 1229 inherited those traits from him.

THAT MONTH I saw a video a visitor had taken of a bull bison walking toward another big bull. August is the bison mating season and the bulls frequently have violent fights over females. No cow bison were nearby but the traveling bull suddenly charged at the other bull and hit him in the side of the head at full speed, perhaps thirty-five miles per hour. It was an instant kill, for he dropped like a rock to the ground. Seeing that video gave me much more respect for wolves that take on the great risk of hunting bison. It also reminded me of the time I saw a cow bison hook a big male Lamar Canyon wolf with a horn and toss him into the air. No wonder wolves prefer to hunt elk rather than bison.

Since seventeen of the eighteen pups born in 2020 had survived through August 2021, as well as all eight 2021 pups, the Junction wolves were now a very big pack. At times we had counts of thirty-two wolves in the rendezvous site. That success rate was more proof of how well the older family members were cooperating to raise, feed, and protect the recent litters of pups. The Junction pack was a well-run, highly efficient organization.

One of the yearlings developed a bad limp that summer. Yellowstone got an email from a person who was worried about reports of a limping wolf and who asked if it could be captured and placed in a wolf sanctuary. I was asked to respond to the person. I wrote back and said that it would be far more stressful to a wild wolf to be captured and taken away from its family than to be allowed to deal with its injury. The wolf eventually completely recovered.

That incident reminded me of a story Doug Smith had told me years earlier. He was doing a collaring operation and wanted to dart some wolves in a remote pack. On spotting the group, the pilot flew alongside a gray female. She was fast but Doug got a dart in her. After landing he examined her and found that she had only three legs. The lower part of a hind leg was missing. We think it was lost when she was caught in a wire snare outside the park. She later formed a new pack and served as their alpha female. If the Park Service had intervened and put her in a sanctuary, it would have deprived the pack from having a very competent leader who had learned to cope well with her missing leg.

ONE DAY I was watching 907 bedded down near all eight pups. The pups moved off into a low area out of her line of sight. 907 got up, walked to a hill that gave her a good view of the pups, and bedded down there. That incident showed what an attentive mother she was.

That tranquil scene of a mother wolf watching over her pups contrasted with political developments outside the park. The Montana Fish and Wildlife Commission sets the rules for hunting in the state. In recent years, the two hunting units (313 and 316) along the northern border of Yellowstone had a limit of one wolf that could be shot in each zone during the hunting season. In 2021 they voted to change that to an unlimited take. The commissioners also allowed snaring, baiting and night hunting of wolves.

Unit 316 is located just five miles north of the Junction pack's den site at Slough Creek. It would take the wolves only an hour or so of traveling to cross the invisible park border and unknowingly enter a zone where they would be shot on sight. The wolf-hunting season opened on September 15.

Two days later, the Junction family naively stepped into Unit 316. Two pups and a yearling were shot and killed. It was a terrible blow to the pack. But the danger to the pack was far from over, for the wolf-hunting season was not scheduled to end until March 15.

We felt very relieved on September 19 when the Junctions were seen in Lamar Valley. I got a count of twenty-two wolves. They moved to the Chalcedony rendezvous site the next day and I saw the six surviving pups. I checked my records and found that I had got a count of twenty-five in the pack on September 13. The count of twenty-two on the nineteenth corresponded with the reported shooting of three pack members. During that time, a few adults came and went from the pack, so we did not know what the count of wolves would settle out to.

That same month, a big male grizzly often hung out with the Junction wolves in their Lamar Valley rendezvous site. The pups quickly got used to his presence and mostly ignored the bear. One day he bedded down with twenty-seven Junction wolves, looking like a member of the pack. When the wolves got up and moved off, the bear followed along. I figured he was counting on the wolves to make a kill so he could get a share of the meat. We began to refer to him as Friend Bear.

I wondered what the grizzly would do if the Junctions got in a fight with a rival pack. I felt he would likely charge at the other wolves, whom he would not have known. The bear was four times the size of the adult wolves, so he would be a prime asset in a battle. That meant the Junctions had a bouncer the size of the Hulk on their side.

In late October, I saw the Junction wolves traveling single file while out on a hunt. At one point, Friend Bear was in the lead position. The entire Junction group followed behind him, like he was the leader of the pack.

That day we saw about a thousand elk in the same area as the wolves and their grizzly companion. The wolves charged at the big herd and soon there was a confusing mix of wolves and elk running every which way. It got very hard for us to follow the various chases and must have been frustrating for the bear, as grizzlies have notoriously bad eyesight.

One gray wolf picked out an elk that seemed slower than the rest. The wolf caught up with its target and pulled her down. Other wolves noticed the commotion. They ran over and helped the first wolf finish off the three-hundred-pound cow. Within a few minutes, all the other wolves had run there. I had not heard the wolves at the carcass howl so I do not know how those packmates knew there was a kill. Maybe they saw ravens flying to the carcass and that tipped them off. Ravens specialize in aerial reconnaissance and are quick to spot a new kill made by wolves.

I saw the grizzly some distance from the action. He seemed confused about exactly where the wolves were, but it looked like he knew a hunt was in process. Bears rely on their noses, not their eyes, so he ran back and forth with his snout to the ground. He must have found a fresh wolf scent trail, for he ran along it right to where the pack was feeding on the elk. The big bear charged in and the wolves scattered.

The wolves had gotten used to him by then, so they soon came back and fed a few feet from him. That bear was smart to stick with the pack, for it was the time of the year when he needed massive amounts of food in order to store enough fat to survive hibernation. That relationship between the wolves and the grizzly continued on through mid-November. Then he disappeared and likely had denned for the winter. I had never seen anything like that before but guessed that in the long history of wolves and grizzlies living in the same habitats, it was not a unique event.

OCTOBER 8 WAS the anniversary of the heroic death of famous Yellowstone wolf 302. I walked to the location where he had died in 2009, at the age of nine, then visited Yellowstone's Heritage and Research Center where they had just finished mounting his skeleton for a display. The tibia bone in one of his hind legs had been broken and had healed at an odd angle. He also had several broken ribs, several broken vertebrae, a broken toe, and a broken bone at the end of his tail, all injuries likely caused by kicks or stomps from elk. His pelvis was out of alignment, and I noticed that some of his teeth were missing. There were signs he had been bitten on the face, mostly likely by other wolves. On top of everything else, 302 suffered from arthritis.

On November 30, we heard that the Junctions had gone west and ended up north of the park border near the town of Gardiner. That area was part of wolf-hunting unit 313. A black Junction adult female was shot there, the fourth pack member lost in the hunt. After her death, the pack usually numbered twenty: ten blacks and ten grays. There were fourteen adults and six surviving pups from 907's litter of eight.

The Wolf Project captured seven wolves in mid-December. Three of them, a yearling and two pups, were from the

Wapiti Lake pack, the group cofounded by wolf 755. After the collars were placed on the wolves, Jeremy stayed nearby to make sure they recovered properly. It got down to minus 5 degrees Fahrenheit (–20 Celsius) in that area and one pup showed the symptoms of hypothermia. Jeremy rubbed the wolf's fur and draped his winter coat over her body. Those things warmed her back to normal. She fully recovered and later rejoined her family.

That positive story was offset by the news that three more wolves had been shot in Unit 313. We found out that two yearlings and a pup from the Junction pack were taken there. That brought the death toll up to seven for the family: three pups, three yearlings, and one adult female. Our despair over the loss of so many Junctions was exacerbated by our knowledge that the pack spent only about 5 percent of their time out of the park. Soon after that, I got a count of seventeen in the pack. Despite the tragic losses, the Junctions were carrying on with their lives. I saw Black Female pinning 907, her longtime rival. Nearby, a group of pups were playing together. 1276 came over and joined in on the games.

THE FIRST SIGN of the 2022 mating season took place on February 2 when I saw 1276 avert her tail to 1048. He sniffed her but walked off. Alpha Black Male then came over and also sniffed her. It was apparently too early in the season for her scent to trigger more of a reaction from the males. That day Black Female pinned 907 once more.

The Junction alpha pair mated on February 15. The next day, Black Female averted her tail to 1048 but Black Male came over and interrupted a potential mating. Soon after that, 907 also averted her tail to 1048. She did it many times and he sniffed her on each occasion but they did not mate that day.

907 repeatedly solicited 1048 to breed with her. I checked my records and saw that when 907 was alpha female she mated with 1048 in 2017 and 2018. All that showed how a subordinate male like 1048 can be a preferred mating partner, even with an alpha female. I later looked at our DNA records of collared wolves and found that 1048 had sired more known Junction pups than alpha male 1047 or any other male.

As I reviewed 1048's life, I realized I had never seen him challenge other males in the pack, despite his formidable size and weight. He seemed satisfied to just be a pack member and help the mother wolves and pups. I wondered if his easygoing personality and record of tending to the pack's females were characteristics that enabled him to be the most successful known breeding male in the pack.

In early 2022 he was the second-ranking male, so if something were to happen to Black Male he would ascend to the top position. That put him in the Bider category. Laurie Lyman once told me that 1048 was a good stabilizer. I thought about that a lot and felt she was right. He seemed to somehow make the pack work together better.

Black Male got interested in 1276 and tried to mate with her but a big bull bison came along and interrupted their romance. During that mating season, I saw Black Female dominate and pin 907 eight times, but she only pinned 1276 once. Meanwhile, Black Male tied with 907 three times, despite Black Female's aggressive dominance over the older female, her aunt. The big male only mated with Black Female once, which suggested he was more drawn to 907 than he was to the alpha female.

Around that time, the big news story was the Russian invasion of Ukraine. There was a fundraising event for Ukraine in Gardiner and I donated money from my wolf books to the cause. My publisher, Greystone Books, told me that a Ukrainian translation of my book *The Rise of Wolf 8* had just been published. 8 was an underdog who had to fight and defeat a much bigger alpha male to protect his family. His story was like how the Ukrainian army was fighting the much larger Russian forces.

BY THE END of the wolf-hunting season in the three states surrounding Yellowstone (Montana, Idaho, and Wyoming), seven of the eight wolf packs based in Yellowstone had lost members and a total of twenty-five park wolves had been shot. The 8 Mile pack lost their alpha female, which was especially tragic for them. The Phantom pack lost at least six members and ceased to exist. Eight Junction wolves were killed: four pups, three yearlings, and one young adult. All four pups had been born to 907, so it was a terrible personal loss for her. I wondered if she would have the same capacity for resilience that I had seen in 755 and 926. A male in the new Rescue Creek pack was also shot, and he would have been from the Junction family. That meant that a total of nine Junctions or former Junctions were lost in the hunt.

The Wolf Project's annual report for 2021 stated that in the wolf hunts from 2009 to 2020 an average of 4.3 wolves were taken in the hunts in the local states. The killing of twenty-five wolves in the 2021–22 hunt was a 481 percent increase over that long-term average. Nineteen park wolves were killed along the northern border of the park in Montana, an area frequented by the Junction family. Four were killed in the Wyoming hunt and two in Idaho. The loss of so many Yellowstone wolves, about 25 percent of the population, set off a huge reaction in the public, especially against the unlimited number of wolves that could be shot when park wolves crossed over the northern border and entered the Montana hunting units.

A strong argument on our side was tourism. A 2023 article in the *Billings Gazette* estimated that visitors coming to Yellowstone created 8,700 jobs in the region. Many of those jobs were directly related to wolf watching, such as wildlife tour guides. Others were auxiliary to that: jobs in local businesses that served Yellowstone visitors, such as motels, restaurants, and gift shops. We estimated that the Yellowstone wolves contributed $82 million to the local economy. Park visitation had skyrocketed to five million people in 2021 and wolves were the number one or number two animals people wanted to see.

I checked the Montana Department of Fish, Wildlife and Parks website during that time and saw that a person could legally shoot up to ten wolves. A license required to kill ten cost $140. That meant the state was placing a value of $14 on each wolf taken. There is another way to estimate the value of a wolf. Since Yellowstone tends to have an average of a hundred wolves and the wolves bring in $82 million to the local communities, you could say each park wolf is worth $820,000. Wolves are far more valuable to the local economy alive than dead.

During the Montana wolf-hunting season, I often went online to the state's website to check how many wolves had been killed along the northern border of the park. The site did not use the words *killed*, *shot*, or *trapped* but employed the euphemism *harvested*. I objected to that, for *harvest* implies that you are eating something taken from nature. Those twenty-five Yellowstone wolves were not "harvested." They were shot or trapped.

I later came across a phrase I had never heard of before: *semantic infiltration*. It is defined as the art of getting your opponents to use your terms rather than their more accurate ones. That appeared to be the strategy behind the state using the term *harvested*.

IN MARCH, AFTER the hunting season ended, we usually had counts of fifteen Junctions: eleven adults and four pups. On the fourteenth of that month, the pack visited their den site at Slough Creek and wolves dug at the entrances of both the Natal Den and the lower Sage Den.

That month we had the 150th anniversary of the creation of Yellowstone National Park, the world's first national park. The idea of protecting the best of our planet caught on throughout the world. There are now over six thousand national parks in nearly one hundred countries.

On April 5, 907 went in and out of the Sage Den, then slipped into the Natal Den, where she spent some time. Two days later, Black Female also went into the Natal Den and stayed inside for two hours. We spotted 907 bedded down with 1048 downhill from the den area during that time. She howled repeatedly as she looked toward the Natal Den. She had used that den many times in the past, but now I wondered if 907 would have her pups somewhere else because of the presence of Black Female at 907's traditional birthing site.

I looked at the Slough den on April 14 and saw Black Female aggressively chasing 907. Both wolves looked very pregnant. A few days later, we saw pregnant 1276 move away from Black Female with a tucked tail at the den area.

The previous year, when Black Female and 1276 lost all their pups, they helped 907 raise hers. But this year it looked like there was a lot of tension between the aggressive alpha and the two other females. It was going to be a difficult and stressful denning season for 907 and 1276, the opposite of the cooperative pattern the previous year.

11

Building an Alliance: 907's Big Gamble

THERE WAS A big development on April 19, 2022. 907 and 1276 relocated to the den 1109 had dug out in the spring of 2020, south of the road just east of Crystal Creek. I suspected 907 had moved there because she was not willing to put up with the dominating Black Female, and 1276 apparently chose to follow her.

That was a fascinating issue for me, for it added an intriguing emotional aspect to the story of 907's decision to den there. We knew 1109 had made the den large, so there was plenty of room for two litters of pups. That left Black Female back in Slough Creek at the Natal Den, with another wolf we called Fourth Mother using the Sage Den.

I checked my records and confirmed that 907 had denned at Slough Creek six times in earlier years. Both she and 1276 must have known it was a risky decision to have their pups away from that site. Back in 2017, 907 had lost all her pups when she had her litter far away from the Slough den, and when 1276 denned west of Slough Creek in 2021, none of her pups survived. The two pregnant females must have felt the risk was worth it.

A few days later, we saw Black Male bedded down near the southern den. That was a significant development, for it indicated that he was going to support 907 and 1276, along with the two mothers at the northern den. Since he mated with 907 multiple times that year, he would be the father of her pups.

ON MAY 2, I saw signs that 1276 was nursing. Right after that sighting, 907 came out of the southern den. She greeted 1276, then the younger mother went into the den. Later both mothers were underground at the same time. They seemed to get along well and later we saw they regularly nursed each other's pups. It looked like they had formed an alliance. The partnership was a huge asset for 907.

Black Female continued to be based at the Natal Den and Fourth Mother was at the Sage Den. At times the alpha female would come downhill and slip into the Sage Den. We frequently saw Black Male up at the northern den site while 1048 was based at the southern den. He repeatedly brought food to the two mothers there who were regularly going in and out of that den.

Pups were seen coming out of the Natal Den on May 6. They were about the size of a kitten. Fourth Mother went into that den without any objection from Black Female. That day ten of the fifteen adults in the pack were at the southern den. It was an early sign that most of the adults were aligned with 907 and 1276.

WE FOUND THAT a location known as Dave's Hill, near the entrance to the Slough Creek road, allowed us to watch both the southern and northern den sites. I spent most of my time there, looking through my spotting scope at one site, then the other. On May 9, a grizzly approached the southern den and five Junctions chased it away. One day a young adult wolf carried off a pup from the southern den, probably intending to play with it. 1276 saw what was happening. She took the pup away from the other wolf and carried it back into the den.

As I watched from the hill, I realized what a smart move it was for 907 to relocate to the south side of the road. The Junction adults were making frequent kills in that section of their territory. It was easy for them to bring meat to the mother wolves and their pups at that site. In contrast, getting to Black Female's den to the north involved dealing with three significant obstacles: crossing the busy park road, then swimming both the Lamar River and Slough Creek during the spring high-water season.

907 had picked a site that intercepted pack members coming from carcasses well before they could reach Black Female's pups to the north. I had always thought that a wolf pack is a support system for mother wolves and their pups, and the conveyor system of adults leaving the southern den and coming back with food for the family was a classic example of exactly that.

When wolves fed the large mob of pups at the southern den, they likely had no leftovers to give to the pups at the northern site. Typically, incoming wolves fed the southern pups, played with them, then took long naps. The feeding and playing sessions would work to bond those young adults to the southern pups.

Another issue that favored 907 was the large number of her offspring in the pack. She was nine years old that spring and had had seven previous litters of pups. Jeremy SunderRaj said 907 had more litters than any other female in the pack, past or present. That made her the most successful mother wolf we ever had in Yellowstone. Her surviving adult sons and daughters were used to helping their mother with her new pups.

All four of the Junction yearlings, who had been born to 907 the previous year, were there and they seemed especially fascinated by the big mob of pups going in and out of the den. The yearlings repeatedly shared food with the pups and played with them. On top of all that, 1048 continued to loyally support 907 and her pups. Black Male visited regularly, and I saw him giving food to 907.

One morning when I was on Dave's Hill, I felt a vibration under my boots and thought a heavy truck was going by. I heard later that day it was a 4.2 earthquake that had occurred twenty-five miles away, near my town, Silver Gate. I had been watching the wolves during the quake and they did not seem to react to it. I had previously seen that adults and pups did not react to thunder. The wolves seemed to mainly pay attention to issues directly related to their survival, rather than things that did not seem threatening.

I had seen other examples of this: When a slow-moving grass fire passed through the Junction den area in August

2016, 911 and the other adults and pups did not seem concerned about the fire and smoke. During 949's decline, when we had a nearly full eclipse of the sun, neither 949 nor the nearby elk and bison seemed to acknowledge that anything was unusual. Perhaps to them it appeared that thick clouds were just covering the sun.

When I was away from the Slough Creek area, a video was shot of four big bull bison approaching the southern den. A group of pups saw them and retreated inside, but one bold black pup stayed in place. The bull and the pup ended up face to face and seemed to touch noses. The pup was probably around five pounds while the bull may have weighed a ton. Soon after that, the four bulls suddenly ran off. We think it was because a gray pup had come out of the den and surprised them.

WE SPENT A lot of time counting pups at the two den areas. By late May, we felt we had at least eleven at the northern den and another ten to the south. We noticed that some pups to the north were smaller than the others. That indicated the ones from Fourth Mother had been born later than Black Female's litter. We later counted seven older pups and four younger ones.

As more days went by, it was clear that far more Junction wolves were rallying around 907 at the southern den site than at the traditional den to the north where the alpha female was denning. They were siding with 907 and getting used to having her as the leader of their portion of the pack.

On May 23, I had my nine thousandth day in the park since the wolf reintroduction back in 1995. I hoped to make it to day ten thousand. I went home later that morning to transcribe my field notes. Laurie called in the afternoon and told me that Fourth Mother had shown up at 907's den carrying a black pup. She put the pup into the den, then slipped in herself. To get there, she must have swum both Slough Creek and the much bigger Lamar River, then crossed the park road. I watched a video of the event, and it showed Fourth Mother carrying the pup by its lower back in her mouth. The pup was limp with its head hanging down, the normal response when a mother carries a pup.

When I went back out to watch the southern den, it looked like the small black pup had already been integrated into the mob of pups born at the site. Then a group of pups nursed on Fourth Mother. That meant she was feeding pups that had been born to 907 and 1276, pups she had likely just met for the first time.

The following morning, I saw Fourth Mother carry another black pup west across the Lamar River Bridge. The river was at a high level and had a swift current. I was impressed that the young wolf had figured out how to take advantage of the bridge so she didn't have to hold the pup above the water when she swam across the river.

When she arrived at the southern den with the pup still in her mouth, several wolves, including 907, ran to her and they all had a friendly greeting. The young mother dropped her pup off at the den entrance, where a bigger pup came out to greet it. We got a count of twelve pups and Fourth Mother was soon nursing a mob of them. That day she carried at least three pups from the northern den to the southern den.

That was an important development. In wolf pack politics, Fourth Mother was essentially giving the pack's alpha female a vote of no confidence and siding with 907's faction. She eventually moved five pups to 907's site, where three of the four Junction mothers were now based.

As we watched the wolves at the southern den, it looked like there was a special partnership between 907 and 1276. We gradually saw that the younger mother seemed to act like an enforcer in the new female hierarchy, pinning any lower-ranking females who looked as though they were getting out of line. That freed 907 to lead other aspects of what the pack needed to do.

By May 28, the only wolves we saw at the northern den were Black Female, Black Male, and two pups, one large and one small. The other thirteen adults and fifteen pups were at the southern den. Big male 1048 seemed to be a favorite with the pups at the southern den. When he bedded down, they liked to climb up on his back.

THERE WAS BIG news on May 29. The three mother wolves, with help from a yearling, were moving the pups from the southern den to an unknown location to the east. 907 had carried off the first pup that way, so the move was apparently her idea. That day fifteen pups were taken east. Meanwhile the alpha pair were still with two pups at the northern den.

The next day, we found most of the Junction wolves at a rendezvous site toward the west end of Lamar Valley. We also saw what looked like several newly dug burrows in the area and figured they were serving as temporary dens for the pups. 907 walked to one hole and looked into it. We saw only a few pups that day. At one point, 907 picked up a pup and put it in one of the new burrows. On later days, we saw far more pups in that area. Now the vast majority of the Junction pack was based even farther away from Black Female.

I checked the Slough den every day that spring. On June 2, I spotted Black Male, Black Female, Fourth Mother, and one larger pup. The three adults left the area, likely going on a hunt. Later that morning, 907 visited the Slough den area. The two alphas and younger mother were still away from the site. The pup went to 907 and tried to nurse on her but her milk had dried up by that time. The pup licked her face. 907 then picked up the pup by the belly and carried it off to the south. When she swam the creek, she held the pup above the water. 907 came out on the far side of the creek and continued south with the pup in her mouth. That direction would take her to the southern den.

Since no adults were attending the lone pup, I wondered if 907 felt it had been abandoned. That would explain why she carried it off. I felt she intended to bring it to Lamar Valley where the fifteen other pups from the southern den were. Three years earlier, Black Female had turned on 907, beat her up, and taken the alpha position from her. Yet now 907 seemed to be showing compassion to a pup born to her rival.

When I looked back at the den area, I saw that Black Female had returned. She sniffed around, then went straight south on what must have been 907's scent trail. I left to look for 907 and returned to Slough Creek when I could not find her at any of the usual road crossing sites. Several wolf watchers stayed at Slough and they saw 907 returning to the spot where she had crossed the creek. She did not have the pup. We thought 907 had probably put it down and it had run out of sight into the thick willows.

907 saw Black Female coming at her and charged forward with a raised tail. The two came together and had a big fight. My friend Audra Conklin filmed the battle and let me look at her footage. The two females bit each other as they stood side by side. At first, they seemed to be evenly matched, then 907 got a good bite on the left side of the black's face. But the younger wolf broke free and attacked 907. 907 once again bit her sister's face. Then the two grappled with each other. Soon after that, 907 broke off from the battle and ran back to the south, toward where the pup was last seen. You could see in the video that both females had blood on their coats.

907 was slightly bigger but Black Female was much younger. I watched Audra's video many times and concluded that Black Female won the fight because she was more aggressive, quicker, and more agile than 907. In addition, since it was her pup, I think she was more motivated to dominate the situation. I remembered that when the two females fought in the spring of 2018, 907 prevailed. She was five years old then, in the prime of her life. Now she was nine, a very old wolf, and had lost two fights in a row (in 2019 when she was badly injured and now in 2022) with her rival. Biologist Taylor Rabe saw 907 soon after the fight and said her wounds looked superficial.

We all have heard of the phrase "survival of the fittest." This phrase has usually been interpreted as meaning the survival of the strongest individuals. In this case, it was Black Female. But with a species like wolves, I came to realize that the fittest animals were the ones with the best social and leadership skills. Despite the outcome of that fight, I felt the support 907 had from most of the pack members could enable her to eventually regain the alpha position. I suspected 907 was thinking a few moves ahead of Black Female. Martin Luther King Jr. once said, "The arc of the moral universe is long, but it bends toward justice." I would say the arc of wolf social life bends toward cooperation.

I later heard that prior to the fight Black Female had swum the creek, found the pup, and tried to get it to swim with her back across the creek but it refused to go into the water. It backed away from the creek. The female brought it back to the water three more times but could not get the pup to swim. 907 had carried the pup across the creek in her mouth but apparently Black Female did not know that trick.

I spotted Black Male with the pup on the south side of the creek. He also tried to get it to swim north with him but the pup continued to balk at going into the water. After that, we lost sight of the pup in the thick willows south of the crossing area. The next morning, the pup was back at the Slough den, so the adult wolves had somehow got it across the creek.

IN THE FOLLOWING days, I watched the new rendezvous site every day. To get a good view of the area, I hiked up a ridge north of the road, then focused my spotting scope on the meadow where the adult wolves and pups were hanging out. On June 3, the day after the fight, I counted thirteen pups there, but there were plenty of spots where additional pups could have been sleeping.

The last known sighting of the black pup at the Slough den was on June 8. The main group of Junction adults and pups continued to be in view in Lamar Valley. It now looked like almost all the Junction wolves but Black Female had sided with 907. Eventually that even included Black Male.

As I thought about how the wolves in the pack ended up supporting 907, the phrase "voted with their feet" came to mind. I say that because after 907 and 1276 chose to den south of the road, nearly all the wolves that were based at the northern den left that site, walked south, and joined 907.

I researched that phrase and found helpful material on Wikipedia. During the 1917 Russian Revolution, the czar's soldiers deserted his army, went over to the revolutionaries, and joined their cause. Vladimir Lenin summed up the incident by saying, "They voted with their feet."

The Wikipedia site states that "foot voting is expressing one's preference through one's actions, by voluntarily... [leaving] a situation one does not like, or [moving] to a situation one regards as more beneficial. People who engage in foot voting are said to 'vote with their feet.'" The site then quotes legal scholar Ilya Somin, who defines foot voting as "the ability of the people to choose the political regime under which they wish to live." That describes exactly what the Junction wolves did regarding the rivalry between 907 and Black Female. They chose to live under 907's regime.

After rereading this section, an intriguing thought came to mind: when the Junction wolves voted with their feet to side with 907, they were essentially functioning as a democracy.

It rained a lot on June 11 and through the following morning, June 12. I noticed that the Lamar River was at a very high level. I went home in the late morning that day. In the afternoon, I looked out the front window of my cabin and saw that a foot of water was flowing west through my driveway. I put on hip boots and drove my car to a high spot at the end of the driveway to keep it out of the water.

I thought beavers had plugged a nearby culvert. If they plug up that pipe with sticks, the town floods and I have to clean out the blockage. But I was wrong about the beavers. Later I got information on events leading up to the June 12 flood. There were several late-spring snowstorms in the mountains east of the park. Then there were three days of hot temperatures on June 8, 9, and 10. After that, we got a lot of rain on the eleventh and twelfth. That rain melted the late snow in the high country far more rapidly than the creeks and rivers could absorb and they overflowed their banks. The flood water flowed west from Silver Gate into the park through Soda Butte Creek and the Lamar River, and finally into the Yellowstone River. It washed out three sections of the park road between my cabin and Lamar Valley as well as the five-mile section between Gardiner and Park Headquarters at Mammoth Hot Springs.

The damaged sections of the park road meant I could not get from my cabin to Lamar Valley. The Park Service closed the entire park because of the road damage. Later the South, East, and West Entrances to the park were reopened but the entrance near Silver Gate and the one at Gardiner were kept closed. It looked like they would remain closed for a long time, perhaps even through the winter. That was a big problem for me regarding my studies of the Junction wolves.

I made two trips into the park in the following weeks but had to come in via the East Entrance, a drive that took more than three hours. The Park Service later announced that they were going to upgrade and pave an old dirt road that ran between Gardiner and Mammoth. During the road work, park employees and wildlife guides would be allowed to go in and out of the park at certain designated times.

I contacted two friends in Gardiner, Nathan Varley and Linda Thurston. They own the Yellowstone Wolf Tracker wildlife tour company and I arranged to work for them. Gardiner is a much bigger town than Silver Gate: it has a population of 893. I got a place to stay in Gardiner and started my guiding job on July 23.

From Gardiner, we could drive east as far as Slough Creek but not farther into Lamar Valley because of a damaged section of the road. That meant I probably would not see the Junction pups or know what was going on between 907 and Black Female for some time. The Wolf Project did flights periodically to check on the wolf packs, and I heard that the Junctions were still based at the west end of Lamar Valley. Jeremy told me that the pack had fifteen adults along with fifteen pups. No wolves were seen at the Slough den area.

I GOT MY first look at the Junctions that summer on August 15. The pack had a bison carcass west of Slough Creek. I saw 907, 1048, 1276, Black Female, and Black Male in the group. The females did not interact, so I was unable to see who was the alpha. This was my first sighting of the Junctions since June 12, the day of the flood.

The Montana Fish and Wildlife Commission held a hearing in Helena, Montana, on August 25. They allowed people to voice their opinions on wolf-hunting regulations and the vast majority testified in favor of greatly lowering the take along the northern border of Yellowstone. Prior to the hearing, people from all over the country wrote letters and emails imploring the commissioners to reduce the killing of park wolves.

At the end of the hearing, the commissioners voted to set a limit of six wolves that could be shot or trapped in that area. The previous year, when there were no limits, nineteen wolves were taken in that zone. We considered it a great victory. The wolf-hunting season would open on September 15 and run to March 15 unless the limit of six wolves was reached before that. For the entire state of Montana, the quota was set at 450 wolves.

12

Leader of the Pack: 907 Prevails

WE GOT A good view of the Junction pack on September 22, 2022. Jeremy SunderRaj and Taylor Rabe found them prior to my arrival and got a count of twenty-two. I saw the pack, including a mob of pups, chasing a bull elk. He ran into the Yellowstone River to escape them. 907 was out in front, leading the chase. She jumped in the water without hesitation and swam after him but had to give up when he got too far ahead of her. This was the first day the Wolf Project crew saw pups traveling a long distance with the adults, and that chase was likely their first hunt. 907 looked strong, vigorous, and decisive as she led the pack.

I had a better view of the wolves the following day. They were south of Slough Creek and going east, back toward Lamar Valley. 907 was leading once again. She was moving fast and had a big piece of meat in her mouth. We got a count of twenty-six wolves with fifteen pups. Black Female was not in the group that day or the previous day and the pack seemed to be doing fine without her.

I later thought about those two sightings of the Junction wolves. In the first one, 907 led the chase of the bull elk; in the second sighting, 907 led the pack back to the pups at the family's rendezvous site in Lamar Valley. That was strong evidence that 907 was the leader of the pack.

Jeremy told me that the Wolf Project counted sixty-four pups throughout the park during the denning season. Now in late August the count of surviving pups was fifty-seven. That was considered a good survival percentage. The success in pup rearing by the Junctions and other Yellowstone packs in 2022 was another example of the ability of wolves to move forward after terrible tragedies. Twenty-five park wolves had been killed in the 2021–22 hunting and trapping season but now more than twice that many pups were being raised.

We had another important sighting of the Junctions later in September. They were traveling through the Slough Creek area. The group included 907, 1276, and Black Female. I noticed that Black Female had a tucked tail and seemed to be fearful of being near the other wolves. I took that as a sign that she had accepted 907 as the alpha and her position as a low-ranking pack member. But to be sure, I needed to see the two females interact with each other.

I HAD BEEN counting the days to October 15. That was the day the road was scheduled to reopen all the way to my town of Silver Gate. I left Gardiner early that morning, drove through the North Entrance Station to Mammoth Hot

Springs, then continued east. We spotted twenty-two Junctions east of Slough Creek. Nearby was a sickly cow bison who was standing by a big bull bison. The wolves approached the cow and nipped at her. The bull charged at the pack a few times, then seemed to lose interest in protecting the cow. The wolves pulled the cow down and finished her off. The Junctions seemed to be making more bison kills than in past years. Jeremy told me that bison now made up about 20 percent of the wolf kills in the park.

After watching the wolves feed for a while, I continued east. I drove through several sections of repaired road, then finally reached my cabin in Silver Gate. I had left it on July 22, so I had been gone a little under three months. Everything was okay with the cabin both inside and outside. The flood had not damaged anything, thanks to a cinder-block foundation that had kept the cabin above the water.

October 22 was a significant day, for that morning I saw 907 dominate Black Female, who kept her tail tucked whenever she was near 907. This made it clear 907 was once again the pack's alpha female. 907 had lost her two previous physical altercations with Black Female. Although she was larger than her rival, she was older and blind in one eye. This time around, instead of confronting Black Female in a physical fight, 907 adopted a Bider strategy of forming a cooperative alliance and waiting for pack members to join her. It was her third reign and it had taken her three and a half years to reclaim the alpha position. Like 755 and 926 before her, 907 had a tremendous capacity for resilience.

After that, we had many sightings of 1276, 907's ally, pinning Black Female. That meant 1276 was now the beta or second-ranking female in the pack and still serving as 907's enforcer. Later I saw 907 approach 1276 with a raised tail and act dominant toward her, but she did not physically bother 1276. It seemed to be more of a ceremonial act than an aggressive behavior. By that time, we were seeing 907 and Black Male doing double scent marking, a confirmation they were the alpha pair.

LATER IN OCTOBER, I went on a weeklong lecture tour to promote my new book *The Alpha Female Wolf*, which was mostly about the 06 Female. I had been invited to join the Explorers Club earlier in the year. I flew to New York City and spoke about 06's life story in the club's headquarters. The following evening, I did a talk at a New York wolf sanctuary called the Wolf Conservation Center. In the middle of my show, the wolves suddenly had a loud group howl.

Later in the tour I flew to Colorado, where I did a talk at the Colorado State University Pueblo. Their sports teams are called the ThunderWolves, so there was a big turnout. When the host finished introducing me, the entire cheerleading team marched into the auditorium chanting my name and waving their pom-poms. That never happened when I was in high school or college. After that, I spoke at the Sundance resort in Utah for the third time.

Doug Smith unexpectedly announced that he would be retiring as head of Yellowstone's Wolf Project in late November. It was a big surprise to all of us. I told him that everything I had accomplished with wolves in Yellowstone was due to him. If he had not hired me to work for the Wolf Project in the spring of 1998, my life would have turned out completely different and my wolf books would not have existed.

BY LATE OCTOBER, it appeared that 1048 had left the Junction pack. Jeremy did a tracking flight on November 6 and spotted him in Pelican Valley, close to some Mollies wolves. On November 10, we saw 1048 traveling with a group of eight wolves west of Slough Creek. The group was mostly Mollies, including their alpha female. 1048 stayed with the pack when they went back to Pelican Valley, which meant he was now living with a whole new set of unrelated females he could breed with. Two other former Junction males who had helped form the Rescue Creek pack, 1272 and an uncollared black, also joined the Mollies. The Wolf Project later determined that the black became the pack's new alpha male.

907 continued to dominate Black Female, usually by chasing and pinning her, but 1276 was more aggressive to the former alpha than 907 was. Although 907 was the alpha female, her style of leadership did not include constant aggression toward lower-ranking wolves. She pinned Black Female on this occasion, and I later saw her briefly pin 1276, but that behavior was moderate compared with other alpha females I have known. Michael Sypniewski, a guide with Yellowstone Wolf Tracker, saw 1276 and other Junction wolves chase Black Female. When they caught up with her, they pulled her down and bit her. After that, 1276 stood over her with a raised tail for several minutes.

That December I saw 1276 approaching the Junction pack when they were bedded down at Slough Creek. She wagged her tail as she greeted several wolves, then lowered it as she approached 907. 907 got up and stood by the other female with her tail raised, a clear sign of her higher ranking. After that, both older females raised their tails as three younger wolves approached them. 907 later romped playfully with

Black Male. He scent marked a site and she marked over the spot.

Later that day, 1276 ended up with a group of nine pups. They ran at a big bull elk and encircled him. The pups took turns approaching the bull. They tried to nip him but stopped short of making contact. He acted confident despite having ten wolves surrounding him. 1276 watched the bull closely and apparently decided he was too strong to mess with. She walked off. The nine pups turned around and followed her. After that incident, I noticed that pups sometimes led the pack when they all were traveling. All of the pups were easily keeping up with the adults. That was a sign that the pups were strong and healthy.

When I got back to Silver Gate, I saw our local bull bison near my cabin. He spent the winters in our town. The big bull would find a promising spot to feed, then sweep his massive head from side to side to clear snow from the underlying grass and feed on it. Our yard was covered with the craters he made. It looked like someone had been digging for buried treasure. As I continued to watch the bull, he would regularly stop and nibble on individual grass stalks that stuck out of the snow. For a one-ton animal, that was not much of a snack. One early morning when I was walking out in the dark to start my car, I heard something stampeding toward me. It was the big bull bison. He ran right past me in the dark. If he had veered a few feet off that angle, I would have been knocked to the ground and trampled.

The official low in Lamar Valley on December 22 was minus 47 (-44 Celsius). A few days later, it was plus 31 (around 0 Celsius), 78 degrees warmer.

EARLY ON THE morning of January 3, 2023, we spotted the Junction pack across the road from the den forest where 926 had been born and where she had raised her pups. 907 was leading the pack east. I saw alpha male Black Male and 1276 in the group, which numbered twenty, including a lot of pups. 907 brought the family to an open area, where they bedded down.

Soon after that, we saw Black Female following the travel route of the other Junction wolves. As she got closer to the bedded wolves, she tucked her tail into a U shape under her body, lowered her head into a subordinate position, and went into a crouch. A gray wolf from the main group ran toward her and Black Female responded by turning around and running away. Soon nineteen wolves were pursuing her. Black Female stopped and turned toward them in a submissive crouch. A young gray greeted her, then a black wolf chased her off. As she ran away, the other wolves lost interest and went back to their bedding spot.

After a while, Black Female approached the pack again, but she was moving slowly and hesitantly as though expecting to be chased or attacked. She ended up bedding down by herself well away from the pack. It looked like she had lost all of her self-confidence and I got the impression that the pack's onetime alpha female could not quite figure out how she had ended up as such a low-ranking adult.

In the late afternoon, Black Female got up, looked toward the wolves to the east, then slowly walked off in the opposite direction. I checked the main group and saw that 907 was up and interacting with Black Male. She did a scent mark and Black Male marked her site. Both wolves wagged their tails at each other, then playfully romped around together side by side. After that, 907 led east and all the other wolves in the group followed her. Apparently intimidated by 907, Black Female maintained a good distance between herself and the other wolves all afternoon.

THE WOLF PROJECT staff captured and radio-collared five Junction wolves on January 5. These included Black Female, who was assigned the number 1382. She weighed 114 pounds, which was much heavier than I'd thought. When 907 was recently recollared, she was 119 pounds. That meant the two females were fairly evenly matched in battle; however, 907 was at a disadvantage when they fought, as she was older and had lost the sight in her left eye, which meant that she could not see Black Female coming at her from her left side. I wondered if that was why 907 had switched from physical altercations to forming a cooperative alliance to get the pack on her side and alienate her rival.

Once she was back in the alpha position, 907 excelled in getting her pack members to work together in feeding and raising pups, hunting prey animals far larger than they were, and defending the family from rival packs. When I recently watched twenty-three members of the Junction pack walking single file up a snowy slope, I recalled a line from *The Lord of the Rings* and modified it to describe 907: one wolf to unite them all.

In late January, we spotted the Junction wolves on top of Specimen Ridge, south of Slough Creek. When my friends Laurie Lyman and Wendy Busch went back out to check on the Junctions later that day, they saw Black Female slowly approaching the pack from the west. Two black wolves from the main group jumped up and ran at her. Black Female took off and ran away with a tucked tail.

A few weeks later, the Junction wolves were in the Hell-roaring Creek area. Black Female was bedded down well away from the other wolves. She seemed to understand she was unwelcome. In late February, I got a report of a young Junction wolf chasing Black Female away from the pack and an elk they had killed.

Black Female appeared to be the lowest-ranking adult in the pack: the omega wolf. Her story reminds me of a line from a 1939 movie called *The Roaring Twenties*. James Cagney played a successful gangster who later in life fell on hard times. He was shot and mortally wounded by a rival. After he died, a cop came along and asked Cagney's girlfriend who he was. She answered, "This is Eddie Bartlett. He used to be a big shot." That was now Black Female's story: she used to be a big shot.

We all worried about the ongoing Montana wolf-hunting and trapping season along the northern border of the park. We got word that dispersing Junction female 1229 had been caught in a trap near Gardiner and shot. She was the fifth wolf taken in that zone and there was a limit of six. Later a Rescue Creek male was shot and that closed the wolf-hunting season. Four of those six wolves were from the park, but none were current members of the Junction pack. We all breathed a sigh of relief.

In the previous wolf-hunting season, eight Junctions had been killed, including four of 907's eight pups. She and the other Junction wolves reacted to that terrible loss by raising fifteen pups in 2022. The pack still had those fifteen pups, along with ten adults.

As I write this in early 2023, 907 will soon turn ten years old. That is triple the average life span of a Yellowstone wolf. She is the oldest known wolf in the park.

In Yellowstone the average number of pups in a litter is 4.5. That suggests that as of the spring of 2023, 907 will have given birth to around forty pups. A chart that veterinarians use lists a ten-year-old dog as being equivalent to a seventy-eight-year-old human, so 907 is in that age range. With 907 once again as pack leader pulling all twenty-five pack members to work together, the pack has entered a new era that could be called Junction United.

AS THE JUNCTION Butte wolves were getting ready for the new denning season, I realized that Junction-descended pups would be born not only in the original pack but also in the Rescue Creek, Shrimp Lake, and Mollies families.

All five adult males in the Rescue pack are Junction wolves, including the alpha male. There are three adult males in the Mollies pack that came from Junction, including their alpha male and 1048. Former Junction wolf 1228 is the alpha female of the Shrimp Lake pack.

That means that four of Yellowstone's ten wolf packs have Junction wolves in them who are breeders. That is compelling proof of the success of the pack and its genetic line.

WHILE WORKING ON this book, I often thought of the differences and similarities of the two main packs I was writing about: the Junction Butte and Lamar Canyon families. Each one had its own unique challenges, but they also had many things in common.

The most important similar events were the killing of their alpha males, 925 and 911, by different groups of Prospect Peak wolves and the later acceptance of some of the males from that pack into the Lamar and Junction packs. The surviving females in the two families employed that same strategy and it enabled their packs to quickly bounce back to normal strength.

Both packs went through several tough years when their pups died from mange, distemper, and other causes. But no matter how many of their pups were lost, the mother wolves never gave in to despair and had litters again in the following springs.

There was continual jockeying for power among the Junction females and frequent turnover in the alpha position. In contrast, once 926 took over the Lamar pack the female hierarchy was stable for years, and there was a peaceful transition when her daughter became the next alpha.

Some events that seemed to be unique in one pack were later somewhat replicated in the other pack. For example, Lamar Canyon alpha male Husky Black left the pack and returned to his birth family. In the Junctions, high-ranking male 890 also left his family. He took on the risk of going into the Mollies territory, where he was accepted into that pack.

The males and females in both packs showed that wolves have different personalities and differing strategies in life. Some seem to be especially ambitious to become alphas, while others do not appear to be obsessed with—or even particularly interested in—advancing in the hierarchy.

The issue that struck me as being most similar was the level of cooperation within each of the two packs. The Lamar wolves did that well, while the Junctions accomplished it despite a lot of rivalries between their female members.

In the end, I saw that wolf families are similar to human families in many ways, especially large ones. Each member has the sometimes difficult task of balancing their desire for personal advancement with the need to cooperate with others in the group.

13

Humans
and Wolves

PRIOR TO MY twenty-eight years of studying wolves in Yellowstone, I watched wolves in Alaska's Denali National Park for fifteen summers. Then I spent one summer in wolf country while working in the Polebridge section of Glacier National Park in northern Montana. What I learned over those forty-four years is that wolves experience a lot of failures and hard times, including deaths of mates and young family members. But the defining trait of their species is their resilience in the face of personal loss.

How did resilience become so characteristic of wolves? For many thousands of years, they had to contend with rival wolf packs and fellow predators such as grizzlies and cougars. In more recent centuries, their biggest enemy has been us.

Originally the species known as the gray wolf lived throughout Europe and Asia, from Ireland to Japan, and in northern Africa. In North America, a large population of gray wolves ranged from the Arctic Ocean in the far north to what is now southern Mexico. They roamed through the Yellowstone area for an estimated fifteen thousand years before the establishment of the park. Except for humans, gray wolves were the most widespread large mammal species on Earth.

Indigenous people coexisted with wolves and had great respect for them. But in Europe and Asia, things eventually changed for the worse. Wolves were considered competitors for wild game animals and threats to livestock. They were all killed off in the British Isles, parts of Asia, including Japan, and much of the European continent. In North America, the early colonists devoted extraordinary efforts to eradicating wolves. I wrote about the history of our efforts to exterminate wolves in my 1995 book *War Against the Wolf*. In the book, I included many colonial and US government documents that recorded the destruction.

Starting in 1630, colonial governments offered cash bounties for turning in dead wolves. The war against the wolf jumped into a much higher gear in 1915, when the US Congress passed a bill that appropriated money for the purpose of "destroying wolves." A federal agency called the US Bureau of Biological Survey was tasked with destroying wolves, along with other predators. In that agency's 1919 annual report for the New Mexico district, there is a passage that reads: "No reasonable argument can be advanced in favor of preservation of any lobo or timber wolves, mountain lions or predatory bears. Their case is simplified by... absolute extermination on the open range."

The government's campaign to kill off wolves even included national parks such as Yellowstone, where the rangers shot and trapped park wolves. They finished that job in 1926, when the last two Yellowstone wolves were killed a mile east of Lamar Valley. Wolves were wiped out in the rest of the lower forty-eight states except for remote parts of northern Minnesota. Because of vast tracts of mostly uninhabited land, larger populations of wolves survived in much of Canada and in Alaska.

For centuries humans did everything they could think of to kill off wolves throughout the world. Perhaps it was partly because of our species' attempt to exterminate them that wolves evolved to be so tough and resilient. That resilience and their cooperative social structure enabled their species to overcome our attempts to kill them off. All that made me think of the famous quote from Friedrich Nietzsche in *Twilight of the Idols*: "Out of life's school of war—what does not kill me makes me stronger."

In the 1960s, attitudes toward predators gradually changed in the United States. Wolves were put on the endangered species list in 1974 and were reintroduced to the northern Rocky Mountain states and Yellowstone in 1995, and later to the southwestern states of New Mexico and Arizona. Then another trait of wolves, the willingness of young adults to take on the risk of dispersing into new country and forming packs there, did the rest.

Wolves from the northern reintroduction dispersed from Montana, Wyoming, and Idaho to Washington, Oregon, and, more recently, California and Colorado. In what is considered the greatest wildlife restoration project in world history, wolves are now back in all the western states except Utah and Nevada, and they may soon disperse there. To state it more simply, in the United States, the species that had sought to exterminate wolves gave them the chance to repopulate the original range where their ancestors once lived. Those wolves seized that opportunity and are successfully accomplishing that mission.

THOSE OF US who live in the Yellowstone area believe in sharing the experience of watching wolves with newcomers. That means we welcome them to the park and let them look at the wolves through our high-powered spotting scopes. I feel that every new person we help will be one more person on the side of wolves. I also hope my books make a difference. Recently, a woman came up to me and said she had bought seven copies of my book on wolf 8 and given them to men she knew who were anti-wolf. All of them read the book and told her that the stories of 8 and 21 had changed their attitudes toward wolves. My mission is to reach as many people as possible and bring them to the side of the wolves.

14

Homecoming

EARLIER IN THIS book, I wrote that there was one last 926 story to tell. Now is the time to share that story.

When I had my 2015 heart operation at Billings Clinic, my dreams of 926 and the Lamar Canyon wolves motivated me to do everything I could to recover quickly. After I returned to the park, I hoped I might have a chance to repay 926 for what she had done for me. At the time, I imagined that would mean writing her life story. That plan changed when a hunter shot 926 in my small town of Silver Gate.

A week or two after her death, I saw the local game warden parked on the outskirts of our town. I went to him and identified myself as a Silver Gate resident and Yellowstone wolf researcher. We talked about 926's shooting and his investigation. The warden told me that after examining the scene of the shooting and questioning the hunter and witnesses, he concluded that it was a legal take. That meant the investigation did not need my sample of 926's blood. I now had to figure out what would be a respectful way to return what was in my freezer to the wild.

I use the word *respectful* because I know that in many cultures blood is regarded as a sacred element. In the Old Testament, there is a passage where Moses is told that blood must be drained from an animal before the meat can be eaten. The specific commandment in Genesis 9:4 states: "You shall not eat flesh with its life, that is, its blood." In a footnote, the editor of my edition of the Bible states that the original Hebrew word, translated in the verse as "life," literally meant "soul." To put that thought in more secular terms, you could say that blood is the essence of an individual's being, both for humans and for animals.

While I was trying to figure out what to do, I thought a lot about 926. I knew her parents and other ancestors, going back to two of her great-grandmothers: wolf 9 and wolf 39. Both were matriarchs of packs brought down from Canada in the 1990s, packs that were named after local features—Rose Creek and Druid Peak—after they settled in Lamar Valley. 926's grandfather was the legendary Druid alpha male wolf 21. Her mother, the 06 Female, was an unstoppable force of nature, and her father, 755, was a wolf with an especially high level of social intelligence. Four of those wolves—39, 21, 06, and 755—raised pups at the Druid den forest in Lamar Valley. That was where 926 was born in 2011 and the place where she raised all her pups.

She had eight males in her life, but I felt her first mate, wolf 925, was the one 926 was most attached to. He died heroically protecting 926 and her pups. Then she converted four of the males who had killed 925 to her side. Later those four either died or left the pack. 926 just moved on and brought three new males into her family. One of them died and a second male left her. The last one was SD and he was with her on her last day of life.

Beyond those details of her epic story, 926 unknowingly touched the lives of tens of thousands of people. Those Yellowstone visitors and local residents saw 926 live out her heroic life and were inspired by her grit and determination to overcome whatever tragedies, setbacks, and hard times befell her. Many more people learned about her exploits through social media posts and video clips.

People often ask me how I deal with tragic events such as the deaths of wolves like 926. I choose to dwell on stories of their valiant, joyful lives, not on the details of their deaths. It is like how Abraham Lincoln is defined not by the way he died but by the way he lived his life.

I still think about 926 nearly every day. I recall seeing her playing with young male SD toward the end of her life. As they romped around together, she seemed to shed the years and revert to being a pup. That was a time when young 926's days were full of affectionate attention from her parents as well as endless play sessions with her four littermates and older siblings.

How could an old wolf like 926 play in such a carefree manner after suffering more heartbreaking tragedies and periods of trauma than any other wolf I have known? The thought came to me that she could put behind her the death of her mother, the loss of three litters of her pups to mange and distemper, and the deaths or disappearance of many males in her life, and live in the moment. I think to her that was the only thing that mattered.

Here is what I learned from watching 926: you do not judge someone by how many times they get knocked down, but by how many times they get right back up. 926 never ran from her problems; she confronted them. 926 did not let circumstances or others dictate her life story. She controlled her destiny. Her story is a celebration of resiliency and empowerment.

In what turned out to be 926's last days, times were good for her. Her daughter had become a competent leader of the pack, the family had a reliable adult male, and they controlled a high-quality territory. But perhaps what meant the most to 926 was that the family had three surviving pups that year. That contrasted with how they had lost all their pups during the three previous years.

In the weeks after we lost 926, I continued to look for her family. In one of those sightings, I saw the Lamar Canyon alpha pair, LT and SD, and two black pups ten miles west of Silver Gate, in a meadow known as Round Prairie. By that time, a female pup had been collared and designated wolf 1197. We saw that the other pup was a male. He became known as Two Dot after a pair of white dots on his chest. Both pups were from the pack's 2018 litter, so they could have had either LT or 926 as a mother. The two adults and pups often played together, a sign they were all in good spirits. We also saw that LT looked pregnant. During that time, the four-member pack ranged from Round Prairie to areas dozens of miles east of the park border.

Late that year, I saw the wolf family feeding on a fresh elk kill in Round Prairie. The two alpha wolves were there, along with Two Dot and two black pups who would have been born after the death of 926. We saw Two Dot scent marking, meaning he was maturing fast. Other sightings of the pack determined the family had four pups. With the three older wolves, that added up to seven. By that time, yearling female 1197 had dispersed and was likely looking for a mate so she could start her own pack.

Soon after that, I spotted the four Lamar Canyon black pups feeding on a carcass just east of Silver Gate. I also saw yearling Two Dot nearby. A friend sighted SD and LT in the area, so that accounted for all seven Lamar Canyon wolves. Six of them were descended from 926. It had been hard on me emotionally when she died a year earlier, but that sighting of her family showed me that she lived on through those six wolves.

I did not want the manner and location of 926's death to be the end of her story. She deserved better than that. Soon after her death, I got an inspiration about what I could do to honor 926's life story. I prepared everything early one morning and went out to Lamar Valley to check on the wolves. I then put on snowshoes and trudged up to the den forest where 926 had been born in 2011 and where she later raised five litters of pups. If any place could be considered her home, it was here.

On the way, I could see the site where 926 had chased the mountain lion up a tree two years earlier. In a different direction, I saw the hillside where I had watched her romping around with her first mate, wolf 925, in 2013.

I stopped by a large Douglas fir at the southern edge of the den forest. That was the tree I saw 926 resting under in the year of my operation, as she watched her new pups playing in the sunlit meadow below. It was the same meadow where 926 had played when she was a carefree pup in the spring of 2011. I felt this was the place where she had her happiest memories. If a wolf like 926 had a favorite spot, I think it would be at the base of that tree.

I took out a plastic bag from my pack. After digging a hole in the snow at the base of the tree, I opened the bag. I gently dropped the blood-soaked snow I had collected where 926 had died into the depression, then brushed fresh powder snow over the spot. 926's genes were in that blood, and I like to think that her soul and essence were as well. I could not bring 926 back to life but I could bring her back home.

In a few months, spring would arrive and the snow around the tree would melt. 926's blood would seep into the ground and the sugar, proteins, and salts in it would be absorbed by the tree's root system. That small amount of blood, her life force, along with moisture and nutrients from the soil, would be carried up through the tree trunk and fuel new growth in the form of wood, needles, and cones. That meant that a little bit of 926 would abide in that tree.

The Douglas fir looked tall enough to have begun life in the 1800s, so it had several more centuries to live. Its upper branches had a commanding view of the den forest where 926 had been born and of her family's territory in Lamar Valley. When the tree died and slowly decomposed, its components, including a little bit of 926, would be taken in by young seedlings who were offspring of the mother tree.

As I snowshoed back down toward the road, I remembered a powerful quote from my friend Scott Frazier. When he spoke at a Native American blessing ceremony on the day the first batch of wolves arrived for the 1995 reintroduction, Scott perfectly summed up the historic significance of that event and also foreshadowed the return of 926 to her home:

"It is good to be part of putting something back, rather than taking something away."

ACKNOWLEDGMENTS

THANKS TO: JEFF Adams, Stacy Allen, Norm Bishop, Taylor Bland, Wendy Busch, Susan and Reve Carberry, Carla Rae Carlson, Kira Cassidy, Lizzie Cato, Melba Coleman, Audra Conklin, Lisa Diekmann, Amanda Evans, Kerry Gunther, Jim Halfpenny, Bill Hamlin, Frank Hogg, Calvin and Lynette Johnston, John Kerr, Bob Landis, Laurie Lyman, Kathie Lynch, Carla Rae and Matt Mathison, Doug McLaughlin, Dave Mech, Cam and Jill Sholly, Deborah Slicer, Doug Smith, Dan Stahler, Erin Stahler, Jeremy Sunder-Raj, Linda Thurston, Blaire Van Valkenburgh, Nathan Varley, Bridgett vonHoldt, Sue Ware, Bob Wayne, Bill Wengeler, Scott Wolff, Tom Zieber.

And a big thank-you to all the wildlife guides in the Yellowstone area!

In appreciation of Yellowstone friends who are no longer with us: Jim Barton, Jeff McIntyre, Don Robertson, Anne Whitbeck.

Special thanks to all the people at my publisher, Greystone Books in Vancouver, British Columbia: my longtime editor, Jane Billinghurst; copy editor Brian Lynch; proofreader Meg Yamamoto; publisher Jen Gauthier; designer Fiona Siu;

marketing director Megan Jones; founding publisher Rob Sanders; and all the other members of the Greystone team who make sure my books get into the hands of readers.

REFERENCES

Barber-Meyer, Shannon M., L. David Mech, and P. J. White. 2008. "Elk calf survival and mortality following wolf restoration in Yellowstone National Park." *Wildlife Monographs* 169: 1–30.

Barrett, Debbie. 2022. "It's Time for FWP to Follow the Law on Elk Management." *Billings Gazette*, July 22, 2022. Note: Barrett cites Montana Department of Fish, Wildlife and Parks data that its target elk population for the state is 92,000, but the current estimated number of elk in the state is 175,000.

Cassidy, Kira A., et al. 2015. "Group composition effects on aggressive interpack interactions of gray wolves in Yellowstone National Park." *Behavioral Ecology* 26: 1352–1360.

Cassidy, Kira A., and Richard T. McIntyre. 2016. "Do gray wolves (*Canis lupus*) support pack mates during aggressive inter-pack interactions?" *Animal Cognition* 5: 939–947.

Duffield, John, Chris Neher, and David Patterson. 2006. *Wolves and People in Yellowstone: Impacts on the Regional Economy.* Bozeman, MT: Yellowstone Park Foundation. Note: In 2006 Dr. Duffield estimated that the presence of wolves in Yellowstone was adding $35.5 million to the local economy. In 2023, because of increased visitation and inflation, the estimated contribution of Yellowstone wolves to the local economy is $82 million.

Harrison, Quinn. "Predictions and Consequences of Dispersal of Gray Wolves (*Canis lupus*) in Yellowstone National Park." Master's thesis, University of Minnesota, 2023.

Hill, Katie. 2023. "How Do Wolves Hunt?" *Outdoor Life*, April 14, 2023. Note: This article includes an interview with Kira Cassidy.

MacNulty, Daniel R., et al. 2012. "Nonlinear effects of group size on the success of wolves hunting elk." *Behavioral Ecology* 23: 75–82.

MacNulty, Daniel R., et al. 2014. "Influence of group size on the success of wolves hunting bison." *PLOS One* 9(11): e112884.

McIntyre, Rick. 1995. *War Against the Wolf*. Stillwater, MN: Voyageur Press.

McIntyre, Rick. 2019. *The Rise of Wolf 8: Witnessing the Triumph of Yellowstone's Underdog*. Vancouver: Greystone Books.

McIntyre, Rick. 2020. *The Reign of Wolf 21: The Saga of Yellowstone's Legendary Druid Pack*. Vancouver: Greystone Books.

McIntyre, Rick. 2021. *The Redemption of Wolf 302: From Renegade to Yellowstone Alpha Male*. Vancouver: Greystone Books.

McIntyre, Rick. 2022. *The Alpha Female Wolf: The Fierce Legacy of Yellowstone's 06*. Vancouver: Greystone Books.

McIntyre, Rick T., and L. David Mech. In press. "Multiple same-sex scent-marking in free ranging gray wolf packs." *Northwestern Naturalist*.

McIntyre, R., J. B. Theberge, M. T. Theberge, and D. W. Smith. 2017. "Behavioral and ecological implications of seasonal variation in the frequency of daytime howling by Yellowstone wolves." *Journal of Mammalogy* 98: 827–834.

Mech, L. David, and Luigi Boitani. 2003. *Wolves: Behavior, Ecology, and Conservation.* Chicago: University of Chicago Press.

Mech, L. D., and R. McIntyre. 2022. "Key observations of flexed-leg urinations in free-ranging gray wolf (*Canis lupus*)." *Canadian Field-Naturalist* 136: 10–12.

Smith, Douglas W., et al. 2010. "Survival of colonizing wolves in the northern Rocky Mountains of the United States, 1982–2004." *Journal of Wildlife Management* 74: 620–634.

Smith, Douglas W., Daniel R. Stahler, and Daniel R. MacNulty, eds. 2020. *Yellowstone Wolves: Science and Discovery in the World's First National Park.* Chicago: University of Chicago Press.

SunderRaj, Jeremy, et al. 2022. "Breeding displacement in gray wolves (*Canis lupus*): Three males usurp breeding position and pup rearing from a neighboring pack in Yellowstone National Park." *PLOS One* 17(11): e0256618.

Online material

Yellowstone Wolf Project annual reports, 2012–2021: www.nps.gov/yell/learn/nature/wolf-reports.htm. (See page 11 of the 2016 annual report for information about wolf 911 and a photo of his skull and jaw.)

INDEX

Agate Creek, 76, 135

Agate Creek wolf pack, 10

—wolf 06. *See under* Lamar Canyon wolf pack

—wolf 113 (alpha male), 10

—wolf 472 (alpha female), 10

Ali, Muhammad, 100–101

Allen, Stacy, 113–14

alpha wolves, 26, 66, 78

altruistic behavior, 73

badgers, 37, 112

Bannock Trail, 69, 70

bark howls, 34, 80. *See also* howls

bears. *See* black bears; grizzly bears

Beartooth wolf pack, 21–22, 47–49, 50

—Husky Black, xxi, 48–49, 50, 203

—SD (Small Dot). *See under* Lamar Canyon wolf pack

—wolf 755: introduction, xxi, xxiii, xxx, 26–27; as alpha male of Wapiti Lake pack, 13–14, 14–15; attempts at new family with Mollies females, 11–13; with Beartooth pack, 21–22, 50; descendants, 2, 23; as Disperser, xxviii, 10, 23–24; displaced from Wapiti Lake pack by Mollies males, 15–18, 18–19, 20; elk hunting, 14, 22; grizzly encounters, 14, 17; loss of wolf 06 and

Lamar Canyon pack, 11; photograph, plate 4; resilience of, xxx, 14, 22–23, 67, 158; social intelligence of, 24, 26–27, 210; wolf 970 and, 76

—wolf 949. *See under* Lamar Canyon wolf pack

—wolf 1292, 23

Biders, xxix, 49, 59, 112–13, 155, 174, 195

bighorn sheep, 52

Billings Gazette, 176

Bishop, Norm, 149

bison: calf rescue, 113–15; encounters with Junction Butte pack, 84, 130, 142, 153, 166, 174, 183, 195; encounters with Lamar Canyon pack, 38–39, 54–55, 56; encounter with Prospect Peak pack, 113; feeding on carcasses of, 37, 53, 54–55, 61, 97, 131, 147, 191; fights between males, 54, 168; Gray Male and, 139; hunting for, 12, 90, 168, 195; injuries from, 43; Mollies pack and, 12, 90; mourning by for dead bison, 153–54; near author's cabin, 198; in Pelican Valley, 12

black bears, 120, 147

Black Female (wolf 1382). *See under* Junction Butte wolf pack